ADVANCES IN ULTRA-LOW EMISSION CONTROL TECHNOLOGIES FOR COAL-FIRED POWER PLANTS

Woodhead Publishing Series in Energy

ADVANCES IN ULTRA-LOW EMISSION CONTROL TECHNOLOGIES FOR COAL-FIRED POWER PLANTS

Edited by

YONGSHENG ZHANG

TAO WANG

WEI-PING PAN

CARLOS E. ROMERO

ELSEVIER

WP
WOODHEAD PUBLISHING
An imprint of Elsevier

Woodhead Publishing is an imprint of Elsevier
The Officers' Mess Business Centre, Royston Road, Duxford, CB22 4QH, United Kingdom
50 Hampshire Street, 5th Floor, Cambridge, MA 02139, United States
The Boulevard, Langford Lane, Kidlington, OX5 1GB, United Kingdom

Library of Congress Cataloging-in-Publication Data
A catalog record for this book is available from the Library of Congress

British Library Cataloguing-in-Publication Data
A catalogue record for this book is available from the British Library

ISBN: 978-0-08-102418-8 (print)
ISBN: 978-0-08-102419-5 (online)

For information on all Woodhead publications
visit our website at https://www.elsevier.com/books-and-journals

Publisher: Katie Hammon
Acquisition Editor: Maria Convey
Editorial Project Manager: Ali Afzal-Khan
Production Project Manager: Maria Bernard
Cover designer: Victoria Pearson

Typeset by SPi Global, India

Contents

Contributors

Carlos E. Romero
Energy Research Center, Lehigh University, Bethlehem, PA, United States

Xingchao Wang
Energy Research Center, Lehigh University, Bethlehem, PA, United States

Tao Wang
School of Energy, Power and Mechanical Engineering, North China Electric Power University, Beijing, China

Wei-ping Pan
School of Energy, Power and Mechanical Engineering, North China Electric Power University, Beijing, China

Yongsheng Zhang
School of Energy, Power and Mechanical Engineering, North China Electric Power University, Beijing, China

Preface

Having a history of nearly 150 years, coal-fired power generation has supported the development of human civilization. Up to now, coal-based power remains as one major source of electricity, even more than 50% of the electricity is produced by coal-fired power plants in some countries. However, the enormous utilization of coal brings forth serious environmental problems. Over the years, researchers and engineers have developed effective methods to improve coal utilization and coal-related pollutant control, while providing a cost-effective path to clean coal power generation. Emissions reduction technologies for applications in coal-fired power plants have significantly reduced the environmental impact of this form of power generation.

China has the largest coal-fired power plant capacity in the world, and become a country with very stringent requirements on air pollution control in connection with coal-fired power plants. Since 2013, China's power generation enterprises have actively advocated and promoted "Ultra-Low Emissions" from coal-fired power plants. "Ultra-Low Emissions" represents strict compliance limits for particulate matter (PM), sulfur dioxide (SO_2), and nitrogen dioxide (NO_x) at 10, 35, and 50 mg/m^3, respectively, subject to a reference condition of oxygen content at 6%. These emissions limits for coal-fired power plant are in accordance with those of gas turbines emissions standards in China's "Air Pollutant Emission Standards for Fossil-Fired Power Plants." These new standards have made the coal-fired power generation industry focus on upgrading and retrofitting conventional pollutant control equipment with new technologies and equipment, which has led to significant advancements in modernizing coal power plants in China and the associated reduction of these pollutants.

What is the technical roadmap for Ultra-Low Emissions for coal-fired power generation, what are the implementation specifics and impact on the plants, what are the pros and cons of treating more than one pollutant, collectively, and what is the cost efficiency of this strict emissions compliance? These are the questions that need to be answered during implementation of Ultra-Low Emissions compliance. The authors of this book have many years of research experience in the field of pollution control from coal-fired power generation. We also have close ties with China's coal-fired power generation industry. We have conducted research and field tests on many coal-fired power units, and held numerous discussions to address low-emissions implementation problems. This work has proven to be helpful to the understanding of Ultra-low Emissions coal-fired plants in China.

This book contains seven chapters. Yongsheng Zhang is the Chief Editor and he has coordinated all the materials for the book. The first, fifth, sixth, and seventh chapters were written by Yongsheng Zhang. The second chapter was written by Tao Wang. The third chapter was written by Carlos. E. Romero and Xingchao Wang. The fourth chapter was written by Tao Wang and Yongsheng Zhang. The third, fifth, and sixth chapters were reviewed by Wei-Ping Pan, and the full text was reviewed by Carlos. E. Romero. Our colleagues and students Chang Song, Zifeng Sui, Jiawei Wang, Yue Peng, Longchun Zhong, and Jiaxin Li also participated in data collection and analysis for this book. Every participant's contribution is greatly appreciated.

CHAPTER ONE

Coal-fired power plants and pollutant emissions

Yongsheng Zhang
School of Energy, Power and Mechanical Engineering, North China Electric Power University, Beijing, China

1.1 Human development depends on energy

Energy acquisition ability is an indicator of human civilization, and it is the foundation of social existence and development. In the book *The Measure of Civilization: How Social Development Decides the Fate of Nations*, Ian Morris claims that energy resources, social organization, war-making capacity, and information technology can be used to measure human civilization [1]. The energy acquisition was broadly defined in this book; it includes the acquisition of food, fuel, and raw materials. Here, fuel refers to the materials, which are used for cooking, heating, refrigeration, kilning, and providing energy for machines such as wood, coal, oil, natural gas, and wind energy, water energy, solar energy, and nuclear energy.

Historically, with the development of human society, more and more energy was demanded by both individuals and society, as shown in Figs. 1.1 [2] and 1.2.

With the development of science and technology and the advancement of society, it is still difficult to quantify whether the energy consumed by humans will reduce or increase. For example, the total population of the world is still increasing now, but the birth rate seems to be getting lower and lower in developed countries. Home office reduces the energy consumed by transportation, but it increases the total energy used for indoor cooling or heating, compared to the situation where the office population density is larger and the energy consumption per capita is lower. Intelligent sharing car reduces the total amount of cars, and autonomous driving is more conducive to the reduction of energy consumption per capita, but the use of a large number of industrial robots may increase energy consumption.

However, the energy consumption per capita of developing countries with a large population is still very low (see Fig. 1.3). For example, the energy consumption per capita of the U.S. is 35 times that of Bangladesh,

Advances in Ultra-low Emission Control Technologies for Coal-Fired Power Plants
https://doi.org/10.1016/B978-0-08-102418-8.00001-2

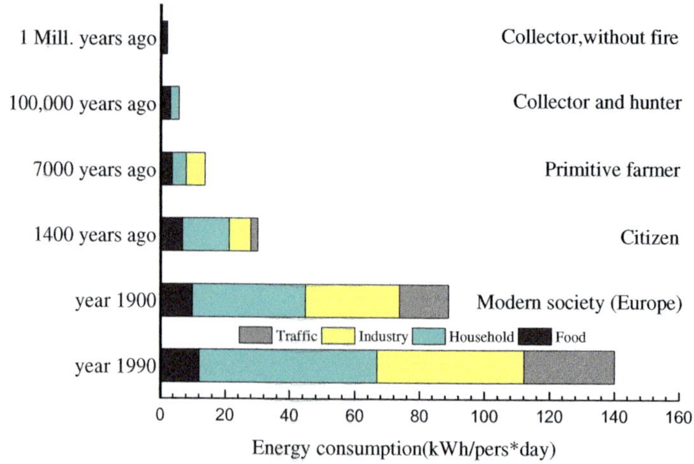

Fig. 1.1 Energy consumption per capita in history. *(Redrawn from Energy and Environment—The European Approach, Klaus Hein, 04/2010 (Workshop Discussion).)*

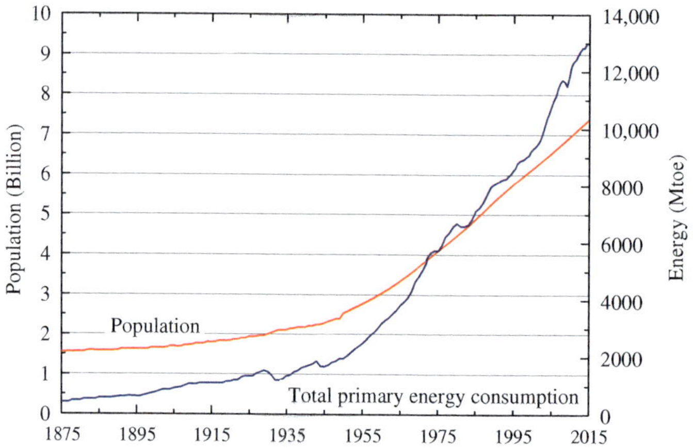

Fig. 1.2 Human energy consumption in different historical periods [3–5].

while Russia is 25 times that of Bangladesh. Although Bangladesh is located in a low-latitude area, it consumes less energy in winter. But the difference between them is still astonishing. Generally speaking, the development degree of the country has a great impact on energy consumption. If the energy consumption per capita of the world reaches the energy consumption per capita of developed countries, the world's total energy consumption will still have a huge potential for increase. Besides, it should be noted that the population of developing countries may increase faster than that of developed countries. For example, by 2040, the population of India is

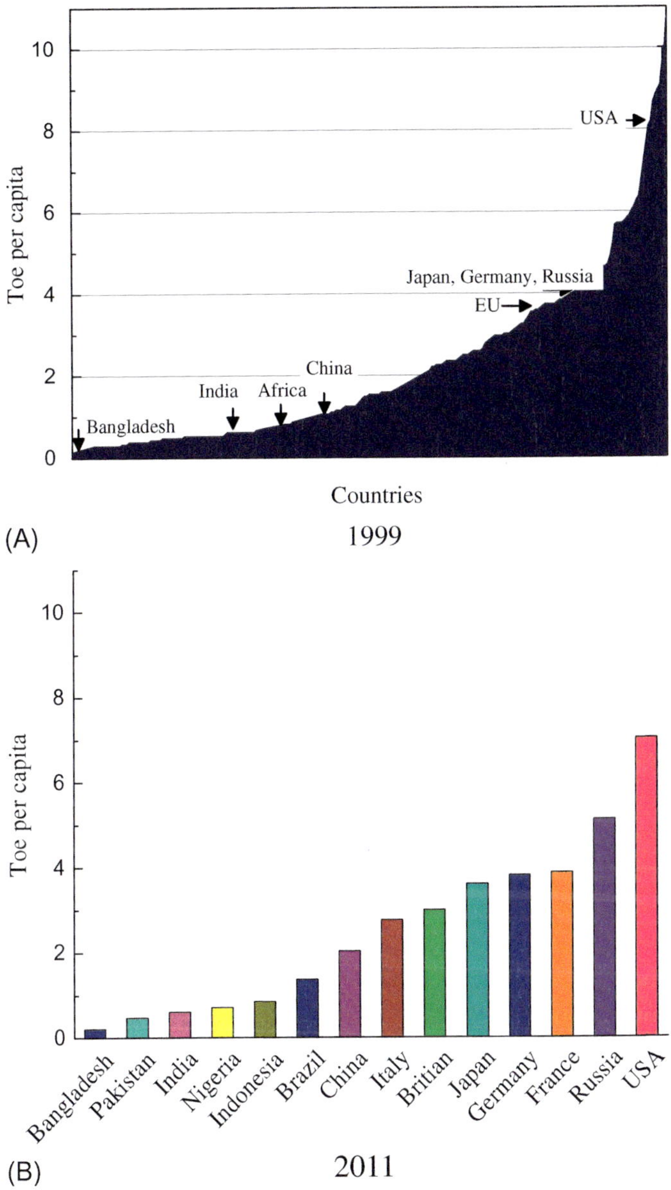

Fig. 1.3 Energy consumption per capita in different countries [6, 7].

expected to reach 1.6 billion. By then, India's electricity demand will quadruple [8]. Also as a populous country in the developing world, although the energy consumption per capita of China has increased in the past decade, it lags behind many developed countries still (see Fig. 1.3A and B). Under the

Table 1.1 Energy consumption estimates by sector of U.S. (Mtoe) [9]

January to January	2017	2016	2015	2014	2013
End-Use Sector					
Residential	59	61	64	70	61
Commercial	43	45	45	47	43
Industrial	67	68	69	70	68
Transportation	55	55	55	54	54
Primary Total	225	228	234	242	226

Table 1.2 Energy consumption sector of China (Mtoe) [10]

Year	2016	2015	2014	2013	2012	2011
Total energy consumption	3051	3009	2981	2918	2815	2709
Agriculture, forestry, animal husbandry, fishery, water conservancy	60	58	57	56	55	54
Industry	2032	2046	2070	2038	1993	1946
Building industry	56	54	53	49	44	42
Transportation, warehousing and postal services	278	268	254	244	228	208
Wholesale, retail and hotel, catering	84	80	76	74	70	64
Others	162	153	141	138	129	118
Life	379	351	330	319	296	277

current energy consumption model, there is still much room for growth in China's energy demand.

In general, the industry consumes more energy in the industrial structure. The proportion of energy consumed by industry in developed countries is low. For example, Table 1.1 shows the energy structure of sector consumption in the United States. The energy consumed by industry is low, accounting for about 30%, which is conducive to the reduction of national overall energy consumption. It can also be found that the proportion of energy consumed by different sectors has not changed much in recent years. For developing countries, the energy consumption structure is different from that of developed countries such as the United States. For example, the current proportion of energy consumption of China's industrial sector is relatively high (66.6% in 2016), and it has decreased year by year (Table 1.2). In the future, the proportion may continue to go down with the development of society and the adjustment of industrial structure, which will lead to a decline in the overall energy consumption of the country. The energy consumption of India's industrial sectors is low, while the energy consumption of its building sector is high (Table 1.3), which may have something to do with its imperfect basic industry. In short, the characteristics of energy consumption of different

Table 1.3 Energy consumption sector of India (Mtoe) [11]

Year	2000	2013	2020	2030	2040	Shares 2013	Shares 2040	2013–2040 Change	2013–2040 CAAGR[a]
Industry	83	185	263	417	572	35%	45%	388	4.3%
Transport	32	75	108	176	280	14%	22%	205	5.0%
Road	28	68	100	165	264	13%	21%	196	5.1%
Buildings	158	214	242	274	299	41%	23%	85	1.2%
Agriculture	15	24	31	43	51	5%	4%	27	2.9%
Nonenergy use[b]	27	29	40	58	72	6%	6%	43	3.4%
Total	315	527	686	968	1275	100%	100%	748	3.3%
Industry, incl. Transformation[c]	111	217	317	507	691	n.a.	n.a.	474	4.4%

[a]Compound average annual growth rate.
[b]Includes petrochemical feedstocks and other nonenergy uses (mainly lubricants and bitumen).
[c]Includes energy demand from blast furnaces and coke ovens (not part of final energy consumption) and petrochemical feedstocks.

countries, especially the countries with large populations, have different impact on the future energy consumption of the world.

1.2 Energy resource that can be used by humans

Currently, the energy resource that can be used by humans includes hydropower, biomass, coal, oil, natural gas, solar energy, wind energy, nuclear fission energy, nuclear fusion energy, geothermal energy, tidal energy, ocean energy, gas hydrate, etc.

Humans use different energy resources in accordance with social conditions and technological levels at different times. Constrained by the technical level, for a long period of time, humans mainly used biomass energy provided by wood, waterwheel, windmills, etc. With the development of the industrial revolution, coal became the main energy resource quickly in the steam era, and even now this proportion is still large in the world energy structure, as shown in Fig. 1.4. Since the 1970s, the social transformation characterized by electrification was called the second social revolution; the oil industry has developed rapidly and has played a significant role in the energy structure. Hydropower also has been widely used because of comparatively simple technology. During the period of the Second World War, the discovery of natural gas resources and the maturity of nuclear fission technology led to a tremendous development of these two forms of energy resources. Since the 21st century, under the environmental pollution

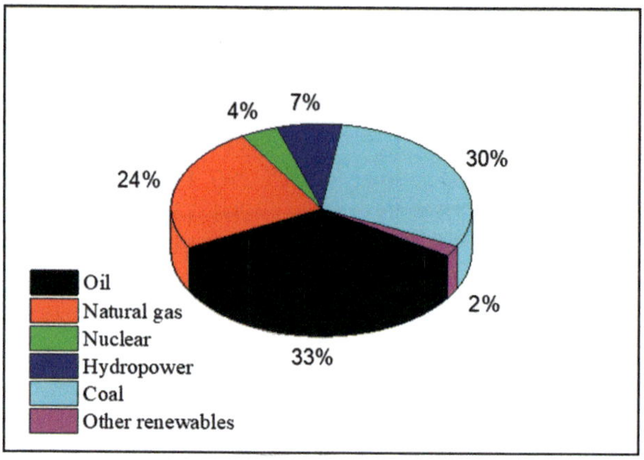

Fig. 1.4 Primary energy demand of the world in 2015 [12]. *(From https://en.wikipedia.org/wiki/File:World_Total_Primary_Energy_Consumption_by_Fuel_(2015).svg.)*

pressure, solar energy and wind energy have been greatly developed due to the demand for renewable energy. Nuclear fusion technology has a very brilliant prospect due to its low carbon, zero pollution, and rich fusion resources on the earth. However, there is still a long way to go for the industrial application of nuclear fusion.

Fig. 1.4 shows the world's primary energy consumption structure in 2015. It can be found that the main primary energy sources are still coal, oil and natural gas. Especially in developing countries such as China and India, the proportion of coal consumption is relatively high, as shown in Figs. 1.5 and 1.6.

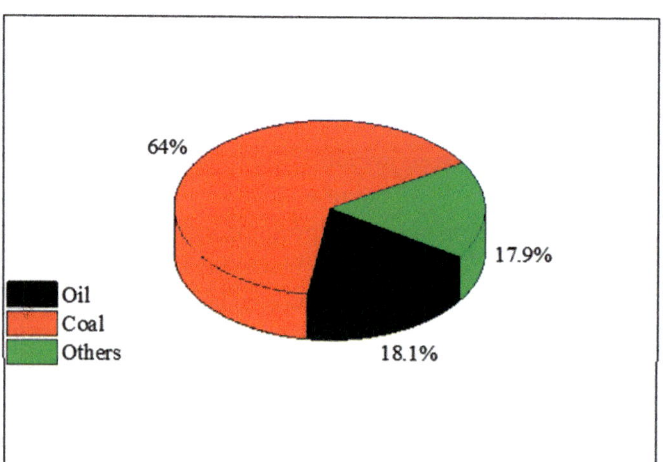

Fig. 1.5 Primary energy demand of China in 2015 [12]. *(Data source: http://www.stats.gov.cn/tjsj/zxfb/201602/t20160229_1323991.html.)*

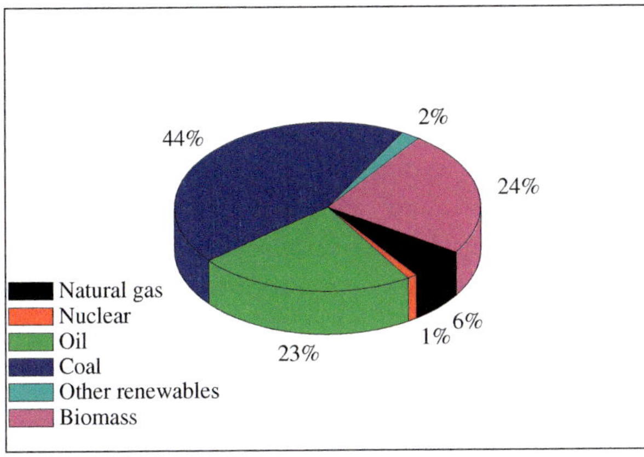

Fig. 1.6 Primary energy demand in India by fuel in 2013 [11]. *(© OECD/IEA 2015 India Energy Outlook: World Energy Outlook Special Report, IEA Publishing. Licence: www.iea.org/t&c.)*

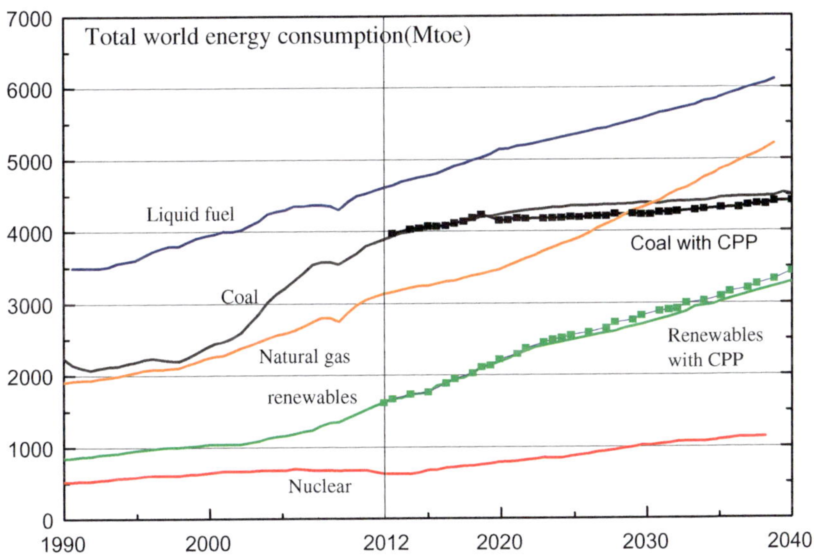

Fig. 1.7 Total world energy consumption by energy source [13].

Based on the actual conditions, it is estimated that coal will still occupy a large proportion in the world's energy structure in the future, as the forecast of the EIA's 2016 report shown in Fig. 1.7. The conclusion is that coal will be the second largest energy source till 2030, and it will be the third largest energy source during the period between 2030 and 2040. In the long run, the total consumption amount of coal may not change too much, but its proportion in energy consumption structure will gradually decrease. Among the methods of using coal, a coal-fired power plant is the most efficient and cleanest way.

1.3 Coal-fired power generation

Electricity is a convenient secondary energy source, which can easily convert primary energy into mechanical energy, heat, cold, light, sound, electromagnetic waves, etc. and can be widely used.

From the data in Fig. 1.8, China, the United States, Japan, Russia, India, and other countries are the major consumers of electricity. However, due to the large population, the installed electricity capacity per capita of China and India is lower than that of the developed countries, such as the United States, Canada, and Germany. Additionally, it is estimated that more than one billion people in Africa, India, and other developing Asian regions are unable to use electricity [14]. There is still a great demand for electricity worldwide.

Humans have tried to convert various primary energy resources, such as hydraulic power, wind, solar, coal, gas, nuclear fission, nuclear fusion, geothermy, etc. into electrical energy in many ways. Across the world, the power generation methods are mainly coal-fired power plant, gas turbine power plant, hydropower plant, and nuclear power. Among them, electric energy converted from coal combustion is the main source of today's electricity. It is estimated that, even by 2040, this type of power plant will still be the main source of electricity.

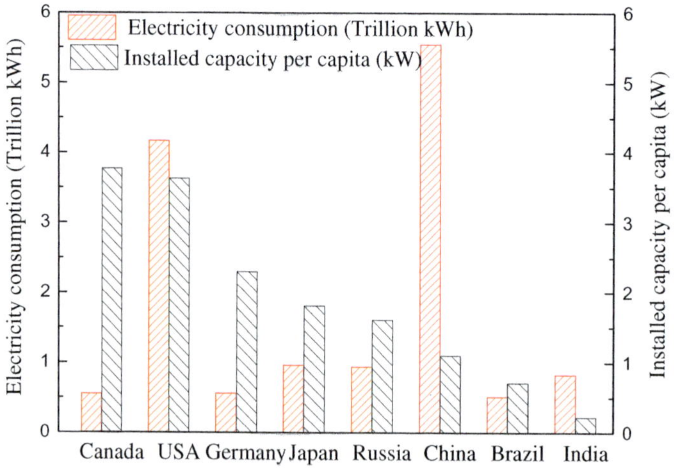

Fig. 1.8 Power consumers and per capita installed capacity in 2015 [12].

In fact, coal does not have many merits other than providing the necessary energy. Why do humans choose coal to generate electricity? The main reason is that coal is a basic energy source that can meet human needs and effectively support the development of human society.

As an energy source capable of supporting human development, it must have the following characteristics:

1) sufficient reserves and stable supply;

2) low conversion costs;

3) mature technology.

Given these factors, we have listed a variety of power generation modes in Fig. 1.9. It is found that the wide application of some power generation modes is restricted variously. For example, the discontinuity of solar energy and wind energy makes it difficult to generate stable power, and high-performance energy storage technology is needed for their further large-scale development; due to the immature technology and high cost, it is predicted that nuclear fusion is unlikely to be the main source of power generation within 30 years; biomass power generation is constrained by raw materials, making it difficult to generate electricity on a large scale.

Therefore, coal-fired power is still irreplaceable for a long period due to its mature technology, low power generation costs, and abundant resources. Especially for developing countries such as China and India, coal power is cheap and cost effective, so it is still the first choice for power generation. In 2016, some coal-fired power plants still were constructed in China, India, Indonesia, Pakistan, Vietnam, United Arab Emirates, Malaysia, Japan, Philippines, South Korea, etc.

1.4 Coal-fired pollution emissions and reduction

1.4.1 Coal-fired power plant emissions

The large-scale use of coal has brought two problems, which influenced widely and lasted for a long time. One is the climate change caused by the emission of CO_2, and another is the air pollution caused by the emission of harmful gases. Due to content limitations, this book only discusses the problem of air pollution.

Among all the existing coal utilization methods, coal-fired power plant is the cleanest method due to high centralization construction, sufficient pollution control funds, and high technology investment. Therefore, the proportion of coal used for power plant in developed countries is very high. The data of 2013 showed that the proportion of coal used for

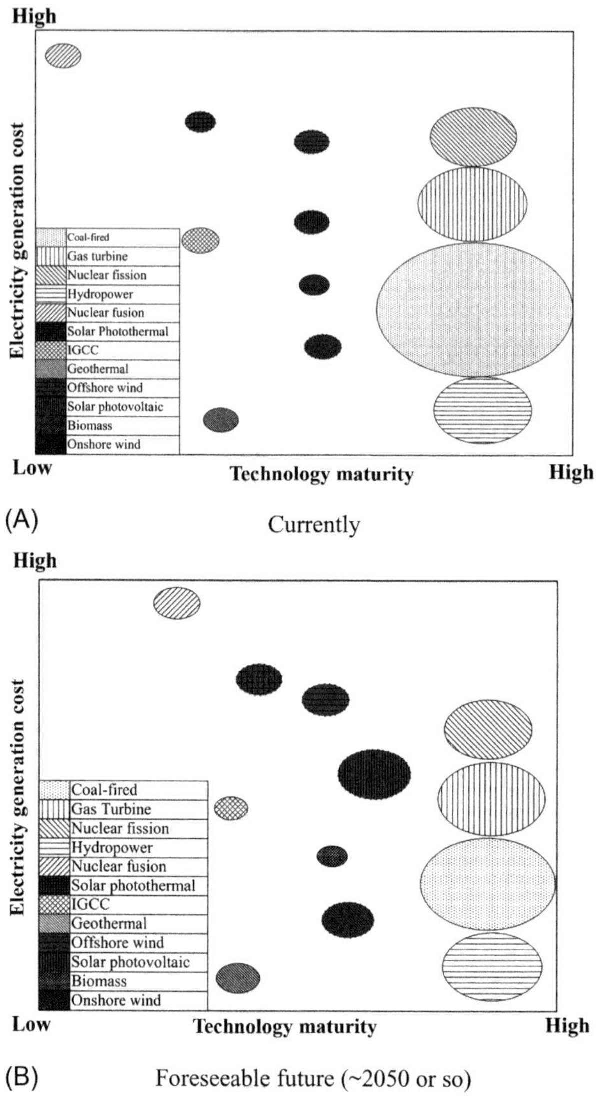

Fig. 1.9 Comparison of different power plant technologies (Note: The area size represents the approximate share of the power plant mode).

power plants in the United States was 92.3% and it was 80.2% in the OECD countries [15]. Developing countries such as China are using more and more coal for power plants. The "Action Plan for Upgrading and Transformation of the Energy Conservation and Emission Reduction of Coal-fired Power (2014–2020)" issued in 2014 stipulates that the proportion of coal consumed for generating electricity should exceed 60% by 2020.

Nevertheless, due to the huge amount of coal consumption, it still has a great impact on the air environment. The pollutants emitted by coal combustion mainly include sulfur oxides (SO_2), nitrogen oxides (NO_x), particle matter (PM, the fine particulate matter ($PM_{2.5}$) of it has more damage to human health). In addition, coal combustion also emits trace metals such as mercury.

In 2015, over 80 million tons (Mt) of SO_2 emissions came from the energy sector, with one-third from the power sector. Over one-quarter of total energy-related SO_2 emissions was discharged from China (22 Mt). India was the next largest source of SO_2 emissions (9 Mt); a development that is spurring increased regulatory efforts to tackle emissions from a coal-dominated power plant [16].

Energy-related emission of NO_x was 107 Mt in 2015, with the power sector accounting for a share of 14%. China (23 Mt) and the United States (13 Mt) account for one-third of global NO_x emissions. Transportation was the largest source of such emissions in many world regions, but China was a notable exception with industry being the largest source. India's NO_x emissions are on an upward path [16].

More than half of global energy-related particulate matter emissions come from the residential sector. China, Africa, and India were the top three regions with high $PM_{2.5}$ emissions. PM emissions are due mainly to incomplete combustion of fuels in households, particularly for cooking (bioenergy), heating (bioenergy and coal), and lighting (kerosene).

Since 2006, the soot, PM (most of them are PM_{10}), SO_2, and NO_x emitted by China's power industry have been decreasing year by year. In 2014, the dust emitted by the power industry accounted for 6% of China's total emissions, and its SO_2 and NO_x emissions accounted for 31% of their total emissions. Nevertheless, the total emissions of power industry are still very large. Based on the statistics of the China Electricity Council, in 2014, the soot, SO_2, and NO_x emitted by the power industry was 976,000 tons, 6.063 million tons, 6.363 million tons, respectively [17].

China and India are the top two consumers of coal in the world. Since 2007, emissions in China have declined by 75%, while those in India have increased by 50%. With these changes, India surpassed China as the world's largest emitter of anthropogenic SO2 in 2016 [18].

The total amount of soot, SO_2, and NO_x emitted by the power industry is calculated, based on the emission concentration and operating situations of units of different sizes. As mercury is a trace element and there is no large-scale on-site monitoring over it, its total emissions data are obtained through indirect methods. For example, Tian et al. estimated that China's emission amount of mercury caused by burning coal was 306 tons in 2007 [19]. Given the mercury content in coal, coal consumption, the use ratio of different furnaces, the removal effect of mercury, and other factors, Huilin et al. concluded that 271.7 tons (147.1–403.6 tons) of mercury were burned by coal-fired power plants in China in 2010, and 101.3 tons (44.0–167.1 tons) of mercury were discharged into the atmosphere [20].

1.4.2 Air pollution and China's actions to reduce emissions

Air pollution is the fourth greatest overall risk factor for human health worldwide. Fine particulate matter is the most damaging to human health, and SO_2, NO_x are also associated with some illnesses.

The latest *State of Global Air 2018* report indicates that 95% of humans are breathing unhealthy air, and air pollution has become one of the biggest risks for human health. In 2016, 4.1 million people worldwide died of heart disease, stroke, lung cancer, acute lung disease, and respiratory infections caused by long-term exposure to $PM_{2.5}$, and half of them are Chinese and Indians. Besides, the air quality gap between developed and developing countries has gradually widened, but China's air quality has been improving [21].

The World Health Organization (WHO) has given the air quality guidelines for concentrations of PM, NO_2, and SO_2 as Table 1.4 shows. The WHO stipulates that the concentration of $PM_{2.5}$ of less than 10 μg/m^3 is safe. In 2016, 95% of the world's population lived in areas with safe concentration of $PM_{2.5}$. For developing countries, the WHO has formulated three different stages of transitional goals, and the first stage sets the concentration limit of 35 μg/m^3. In China, 58% of the population lives in the areas with the concentration of $PM_{2.5}$ exceeding this limit.

The report of *State of Global Air 2018* analyzes China's main sources of air pollution. According to the report, in China, coal burning for industry,

Table 1.4 WHO air quality guidelines for concentrations of PM, NO_2, and SO_2

Mean	$PM_{2.5}$ (μg/m^3)	PM_{10} (μg/m^3)	NO_2 (μg/m^3)	SO_2 (μg/m^3)
Annual	10	20	40	–
24 h	25	50	–	20
1 h	–	–	200	–
10 min	–	–	–	500

power plant, and heating is the main source of $PM_{2.5}$, accounting for 40% of the total $PM_{2.5}$. 155 thousand people died of diseases caused by industrial coal-burning pollution.

The northern, western, and middle eastern regions of the African continent, such as Egypt and Nigeria, have the highest population-weighted annual mean $PM_{2.5}$. Natural factors such as wind and sand are the main reasons for the high concentration of $PM_{2.5}$ in these regions. However, human factors cannot be ignored. In countries such as Niger, Nigeria, and Cameroon, large amounts of indoor solid fuel combustion, as well as outdoor farmland and forest burning exacerbate air pollution.

The air pollution in South Asia is second only to these mentioned regions, and its sources of air pollution are more complex, including municipal solid fuel combustion, coal-fired power plants, agricultural and other outdoor incineration, plus industrial and traffic pollution. In this region, the concentration of population-weighted annual mean $PM_{2.5}$ is 101 $\mu g/m^3$ in Bangladesh, 78 $\mu g/m^3$ in Nepal, and 76 $\mu g/m^3$ in India and Pakistan [21].

Compared with 2010, the global concentration of population-weighted average yearly $PM_{2.5}$ in 2016 increased by 18%. Since 2010, the deterioration of air quality in densely populated areas such as India, Bangladesh, and Pakistan has intensified. The good news is that since 2010, China's population-weighted $PM_{2.5}$ concentration has stabilized and has a small downward trend. In 2016, it approached the world average [21].

Fig. 1.10 shows the average annual emissions of China's 74 key cities from 2013 to 2017. It can be found that the concentration of pollutants

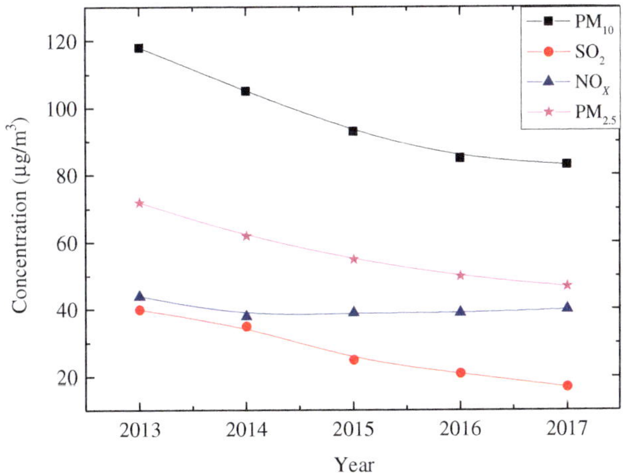

Fig. 1.10 Air pollution of 74 typical cities in China [22].

has been declining over the past 5 years, and the concentration of PM_{10}, $PM_{2.5}$, SO_2, and NO_x in 2017 was 30%, 35%, 58%, and 9% lower than that in 2013 [22].

On a larger scale, the average concentration of PM_{10} in 338 cities in China decreased by 22.7% compared with 2013. The average concentration of $PM_{2.5}$ in the heavily polluted Beijing-Tianjin-Hebei Urban Agglomeration, Yangtze River Delta, and Pearl River Delta decreased by 39.6%, 34.3%, and 27.7%, respectively. The average concentration of $PM_{2.5}$ in Beijing decreased from 89.5 $\mu g/m^3$ in 2013 to 58 $\mu g/m^3$ in 2017. Although there is still a long way to satisfy the standards stipulated by WHO, China's short-term emission reduction results are commendable.

In 2017, China implemented the *Action Plan for Air Pollution Prevention and Control*. The main actions include: ultra-low emission upgrading of the coal-fired power plants with an accumulative number of 700 million kW (accounting for about 18% of the installed capacity of coal-fired power plants in the world); the basic completion of the elimination of small coal-fired boilers in urban built-up areas in the prefecture level and above cities, the cumulative elimination of more than 200 thousand small coal-fired boilers below 10 T/h in urban built-up areas, the implementation of national V motor vehicle emission standards and oil standards nationwide; the basic completion of the elimination of yellow-label vehicles, and the production of more than 1.8 million new energy vehicles; the implementation of the ship emission control zone plan; the initiation of atmospheric heavy pollution causes and governance projects; the implementation of comprehensive control of air pollution in autumn and winter in BeijingTianjin, Hebei Region, and surrounding areas; the renovation of 62 thousand polluting enterprises, the completion of the annual task of replacing coal with gas and electricity, the reduction of the consumption of scattered coal by about 10 million tons.

1.4.3 Ultra-low emission of coal-fired power plant

This book focuses on "ultra-low emission" of coal-fired power plants that have significant impacts on China's air pollutant emissions reduction. This is unlike the previous emission reduction mode, in which the national environmental protection department first formulates standards and then the power plants implement the emission reduction as required. Ultra-low emission is the clean action of coal-fired power plants initiated by China's power generation enterprises in 2013.

Since 2013, China's power generation enterprises have actively advocated and promoted "near zero emission" or "ultra-low emission" of coal-fired power plants [23], through the upgrading of control devices of normal pollutants such as dust, SO_2, NO_x, etc. and the adaptation of new technology and equipment, to achieve the reduction of these pollutants, which has been gradually endorsed and promoted by the government and society. In September 2014, the National Development and Reform Commission, the Ministry of Environmental Protection, and the National Energy Administration issued the *Action Plan for Upgrading and Transformation of the Energy Conservation and Emission Reduction of Coal-fired Power (2014–2020)*, which put forward requirements of "ultra-low emission" on coal-fired power plants that were stricter than the national thermal power plant emission standards. The Action Plan stipulates that the emission concentrations of PM, SO_2, and NO_x shall not be higher than 10, 35, and 50 mg/m^3, respectively, under the condition of 6% reference oxygen concentration, and the emission concentration limit shall be in line with the emission standard of natural gas turbine units of China's GB 13223-2011 "Air Pollutant Emission Standards for Thermal Power Plants." Based on the implementation effect, the newly built or modified ultra-low emission power plants have achieved the expected targets [23–25].

According to incomplete statistics, among the major power generation enterprises in China, by the end of 2016, China Guodian Corporation's 121 units and 52.21 million kilowatts of coal-fired power units achieved ultra-low emissions, accounting for 52.6% of the total installed capacity of coal-fired units. China Huaneng Group completed a total of ultra-low emission retrofit of 69.21 million kilowatts of units, accounting for 59% of coal-fired installed capacity. China Datang Corporation completed the ultra-low emission retrofit of 88 units, and the cumulatively implemented the ultra-low emission of 157 coal-fired power units with a capacity of 64.45 million kilowatts, accounting for 67.8% of the capacity of coal-fired power units in service. China Huadian Corporation increased 34.18 million kilowatts of ultra-low-emission units throughout the year, cumulatively reaching 45.32 million kilowatts, accounting for 51% of coal-fired installed capacity; Shenhua Group Corporation's 75 units achieved 39.76 million kilowatts of ultra-low emission, accounting for 55% of coal-fired power plant installed capacity. Appendix summarizes the emissions of ultra-low emission units of some of China's power generation companies.

According to the data of the Desulfurization and Denitration Committee of China Association of Environmental Protection Industry. In 2016, the capacity of the flue gas desulfurization units of the newly built thermal power plant was about 50 million kilowatts, and the capacity of the flue gas denitration units of the thermal power plant was about 90 million kilowatts. According to the statistics of the committee, as of the end of 2016, the capacity of flue gas desulfurization units of coal-fired power plants nationwide reached 848 million kilowatts, accounting for 80.5% and 90.0% of China's unit capacity of thermal power and coal-fired generating, respectively. The capacity of the flue gas denitration unit of the coal-fired power plant reached about 864 million kilowatts, accounting for 82% and 91.7% of China's unit capacity of thermal power and coal-fired power plant, respectively.

In just a few years, China's coal-fired power plants have achieved remarkable achievements in pollution reduction. According to statistics, after the implementation of ultra-low emission of coal-fired power in China in 2015, the emission of dust, SO2, and NO_x decreased by 93.3%, 85.2%, and 82.0%, respectively, compared with the previous peak emissions [26].

In 2016, the emissions of dust, SO2, and NO_x of thermal power generation per unit decreased to 0.08, 0.39, and 0.36 g, respectively, which were the lowest emission values in the world; from 1979 to 2016, the thermal power plant increased by 17.5 times, but the dust emissions fell by 94% compared with the peak emissions of 6 million tons, SO2 emissions fell by 87% compared with the peak emissions of 13.5 million tons, and NO_x emissions fell by 85% compared with the peak emissions of 10 million tons [27].

China's coal-fired power plant and thermal power plant in 2015 was 2.4 times and 1.5 times that of the United States, and the total annual emissions of PM, SO_2, and NO_x of China (4.2 million tons) were basically the same as those in the United States (4.37 million tons). The detailed data for 1990~2016 are shown in Fig. 1.11A–C [27].

1.5 Main contents of this book

Ultra-low emission is the air pollutant control behavior of target-oriented coal-fired power plants. The coal-fired power units involved include both coal-fired units in operation and newly built coal-fired units. Due to the diversity of natural environment and coal quality, China devised many pollutant control technologies for various kinds of coal-fired units. The ultra-low-emission technology has been upgraded, based on the

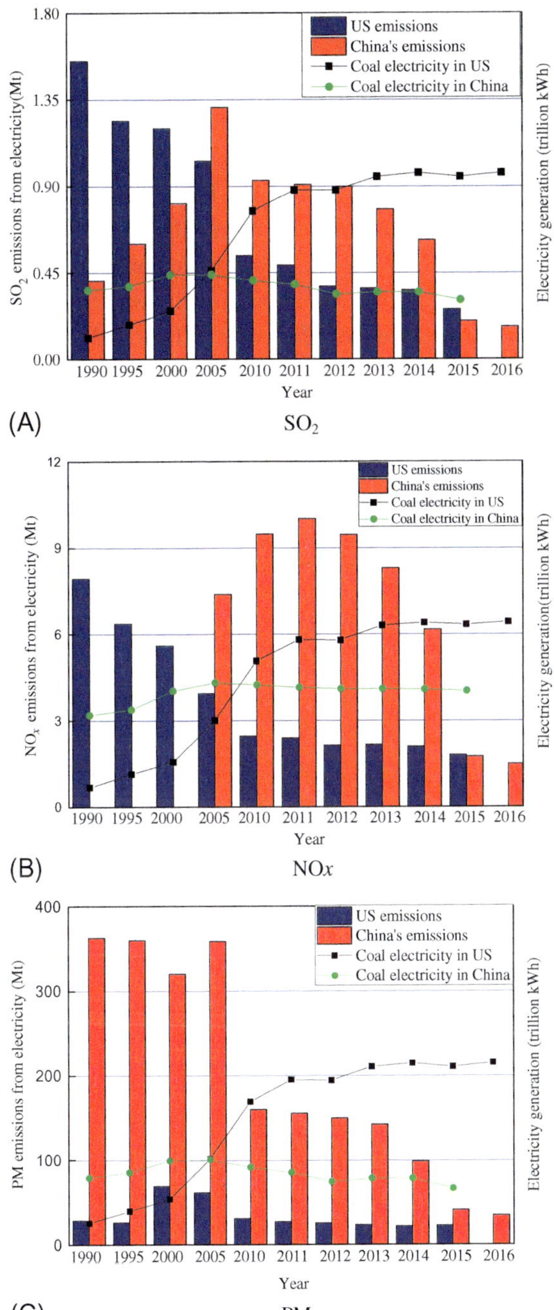

Fig. 1.11 Comparison of power pollutant emissions of China and the US [27].

existing Air Pollution Control Devices (APCDs), and practice has proved that it can be achieved through a variety of technical roadmaps.

This book will discuss these technical roadmaps. Based on the on-site running and monitoring data of coal-fired units, detailed explanations on the pollution emissions of China's typical ultra-low-emission units are given. The pollutants mentioned in this book are mainly the PM, SO_2, and NO_x that can be obtained by the monitoring instruments of coal-fired units. Additionally, the emission characteristics of trace substances such as $PM_{2.5}$ and mercury in ultra-low-emission units will be introduced.

With respect to the layout of this book, Chapter 2 introduces the emission standards of PM, SO_2, NO_x, and mercury pollution from coal-fired power plants in China, the United States, Europe etc. and discusses the technical requirements of China's ultra-low emission and the effect of reducing pollutants in China.

Chapter 3 introduces the key technologies of PM, SO_2, and NO_x emission control and proposes some technology roadmaps of ultra-low emission coal-fired power plants.

Chapter 4 introduces the practical effect of the ultra-low emission project on different technology roadmaps coal-fired units in China. The characteristics of pollutants emission before and after the upgrade of ultra-low emission will be compared.

In Chapter 5, the influence of different pollutant control units on emission reduction of fine particulate matters in the ultra-low emission power plants will be discussed. The influence of different pollutant control units such as wet electrostatic precipitator, desulfurization equipment, and electrostatic precipitator on $PM_{2.5}$, PM_{10}, and fine particles precursor SO_3 will be analyzed.

Chapter 6 discusses the characteristics of mercury and other heavy metal emission from ultra-low emission power plants, discusses the effects of denitrification, desulfurization, and dedusting units on heavy metal emissions, and compares mercury emissions before and after ultra-low emission upgraded.

In Chapter 7, the technical economics of ultra-low-emission power plants will be analyzed.

Appendix

Table A.1 Statistical of ultra-low emission coal-fired power plants

Power generation companies		Province	Power plant	Unit	Capacity (MW)	Acceptance test	Pollutant(mg/m^3)		
							PM	SO_2	NO_x
1		Tianjin	Dagang	1	328.5	2014.2	2.8	13	32
2		Tianjin	Dagang	2	328.5	2013.11	3.33	18	33.67
3		Tianjin	Dagang	3	328.5	2014.5	3.53	9.33	33.33
4		Tianjin	Dagang	4	328.5	2014.3	2.97	17	26.67
5		Tianjin	Panshan	1	530	2015.5	2.36	4.71	36.77
6		Tianjin	Panshan	2	530	2015.12	2.1	4.65	34.48
7	Shenhua Group	Hebei	Sanhe	1	350	2014.7	5	9	35
8		Hebei	Sanhe	2	350	2014.11	3	10	25
9		Hebei	Sanhe	3	300	2015.11	2	12	22
11		Hebei	Sanhe	4	300	2015.7	0.23	5.9	20
12		Hebei	Dingzhou	1	600	2015.12	0.74	8	22
13		Hebei	Dingzhou	2	600	2016.3	0.9	9	29
14		Hebei	Dingzhou	3	660	2014.12	2	6	17
15		Hebei	Dingzhou	4	660	2015.1	2	7	21
16		Hebei	Cangdong	1	600	2016.2	3	9	23
17		Hebei	Cangdong	2	600	2016.3	2	3	20
18		Hebei	Cangdong	3	660	2015.11	4	15	26
19		Hebei	Cangdong	4	660	2015.10	2	10	19
20		Hebei	Qinhuangdao	1	215	2015.7	2.23	8.16	26.5
21		Hebei	Qinhuangdao	2	215	2015.7	2.23	8.16	26.5
22		Hebei	Qinhuangdao	3	320	2014.12	1.5	*	8
23		Hebei	Qinhuangdao	4	320	2015.4	1.6	3	24
24		Inner Mongolia	Hubei	1	600	2016.8	4.9	<15	25
25		Inner Mongolia	Hubei	2	600	2016.8	5	15	20
26		Inner Mongolia	Zhungeer	3	330	2015.9	4.1	21	10.6
27		Inner Mongolia	Zhungeer	4	330	2015.11	3	10.23	25.05
28		Ningxia	Yuanyanghu	2	660	2014.8	4.5	30.4	24.8

Continued

Table A.1 Statistical of ultra-low emission coal-fired power plants—cont'd

	Power generation companies	Province	Power plant	Unit	Capacity (MW)	Acceptance test	Pollutant(mg/m^3) PM	SO_2	NO_x
29	Shenhua Group	Ningxia	Ningdong	1	330	2015.11	4.45	21	30.9
30		Ningxia	Ningdong	2	330	2015.11	3.8	29.2	41.2
31		Liaoning	Suizhong	1	880	2015.2	5	16	27
32		Liaoning	Suizhong	2	880	2014.8	5	30	44
33		Liaoning	Suizhong	3	1000	2015.9	4.42	19	46
34		Shanxi	Hequ	3	600	2016.1	2.4–4.7	11.2–22.6	13.4–38.8
35		Shanxi	Hequ	4	600	2016.6	1.59–4.24	17.7–23.2	12.8–39.3
36		Henan	Jiaozuo	1	660	2014.8	<5	<35	<50
37		Henan	Jiaozuo	2	660	2016.7	<5	<35	<50
38		Henan	Mengjin	1	600	2015.5	3.5	18	36
39		Henan	Mengjin	2	600	2015.4	3.77	12	40
40		Shandong	Shouguang	1	1000	2016.7	<1	2	18
41		Shandong	Shouguang	2	1000	2016.11	<1	<2	16
42		Chongqing	Wanzhou	2	1000	2015.9	2.92	19.9	32.2
43		Jiangsu	Chenjiagang	2	660	2016.6	<2.9	<10	<20
44		Jiangsu	Taicang	7	630	2015.11	0.3～0.4	3～8	15～19
45		Anhui	Maanshan	1	330	2015.12	0.99	5.7	22
46		Anhui	Maanshan	2	330	2016.1	0.91	6	18
47		Anhui	Maanshan	3	330	2015.12	1.06	21	46
48		Anhui	Maanshan	4	330	2016.6	0.98	6	32
49		Anhui	Jiuhua	1	320	2016.6	2.1	8.3	11.9
50		Anhui	Jiuhua	2	320	2015.8	2.35	3.5	13.25
51		Zhejiang	Ninghai	2	600	2016.3	1	3	27
52		Zhejiang	Ninghai	4	600	2015.12	<1	6	17
53		Zhejiang	Zhoushan	3	300	2015.12	0.53	2.87	23.9

54	Shenhua Group	Zhejiang	Zhoushan	4	350	2014.6	2.46	2.76	19.8
55		Guagndong	Huizhou	1	330	2014.12	1.4	8	18
56		Guagndong	Huizhou	2	330	2016.1	0.8	9	18
57	Shenhua Group	Fujiang	Yanshi	5	300	2015.12	4.4	28	41
58		Hebei	Zhangjiakou	3	320	2015.12	<10	<35	<50
59		Hebei	Zhangjiakou	4	320	2015.12	<10	<35	<50
60		Hebei	Zhangjiakou	5	320	2015.12	<10	<35	<50
61		Hebei	Zhangjiakou	6	320	2015.12	<10	<35	<50
62		Hebei	Zhangjiakou	8	320	2015.12	<10	<35	<50
63	China Datang Corporation	Hebei	Qingyuan	1	330	2015.10	2.3	20.1	34.1
64		Hebei	Qingyuan	2	330	2015.11	2.3	20.1	34.1
65		Shanxi	Lingfen	1	300	2015.12	3	9	26.8
66		Shanxi	Baqiao	2	330	2016.6	<5	<35	40
67		Henan	Sanmengxia	4	600	2016.2	<5	<35	<50
68		Jiangsu	Nanjing	1	660	2015.6	2.43	26.91	44.56
69		Jiangsu	Nanjing	2	660	2014.11	4.9	29	35
70		Hebei	Longshan	1	600	2015.6	<10	<35	<50
71		Shanxi	Datong No.2	10	660	2016.5	2	18	36
72	China Guodian Corporation	Shandong	Liaocheng	1	15	2015.6	<5	<35	<100
73		Shandong	Hezhe	5	300	2015.12	<5	<35	<50
74		Shandong	Hezhe	6	300	2016.5	<10	<35	<50
75		Gansu	Fanping	2	330	2015.11	7.02	20.99	73.83
76		Jiangsu	Taizhou	1	1000	2015.4	1.75	16.71	41.16
77		Tianjin	Beijiang	1	1000	2015.4	2.1	6.8	39.1
78		Tianjin	Beijiang	2	1000	2015.11	<20	<100	<100
79		Guangxi	Beibuwan	1	320	2013.12	*	*	<140
80		Guangxi	Beibuwan	2	320	2008.12	<20	<100	<100

Continued

Table A.1 Statistical of ultra-low emission coal-fired power plants—cont'd

	Power generation companies	Province	Power plant	Unit	Capacity (MW)	Acceptance test	Pollutant(mg/m^3)		
							PM	SO_2	NO_x
81	State Power Investment Corporation	Inner Mongolia	Tongliao	5	600	2016.6	<5	<35	<50
82		Henan	Yuxin	6	330	2016.7	<3.58	<16.83	<31.75
83		Henan	Pingdingshan	1	1000	2015.9	3.83–4.05	6–16.8	20.2–23.7
84		Henan	Pingdingshan	2	1000	2015.12	<10	<30	<50
85		Anhui	Pingwei	5	1000	2015.5	<10	<35	<50
86		Henan	Yaomeng	3	300	2016.6	<5	<35	<50
87	China Huadian Corporation	Hebei	Yuhua	1	300	2014.9	4	7	27
88		Hebei	Yuhua	2	300	2014.12	4	29	29
89		Tianjin	Junliangcheng	9	350	2014.11	3.28	<15	24
90		Tianjin	Junliangcheng	10	359	2014.11	3.59	<15	30.5
91		Anhui	Lu'an	4	660	2016.6	3.5	17	40
92		Shandong	Shiliquan	5	660	2015.7	1	13	29
93		Jiangsu	Jinling	1	1000	2015.1	1.7–2.3	1–9	10–12
94		Heilongjiang	Yichun	2	350	2015.12	<10	<35	<50
95		Hunan	Yueyang	6	600	2015.7	<10	<35	<50
96		Shandong	Rizhao	1	350	2015.11	<20	<50	<100
97	China Huaneng Group	Shandong	Rizhao	2	350	2015.11	<30	<50	<100
98		Shandong	Rizhao	3	680	2015.11	<30	<50	<100
99		Shandong	Rizhao	4	680	2015.11	1.8	13.5	35.5
100		Zhejiang	Changxin	1	660	2014.12	2.56	13.67	37.12
101		Zhejiang	Changxin	2	660	2014.12	2.15	17.57	14.8
102	China Resources (Holdings)	Jiangsu	Nanjing	3	300	2016.4	<5	<35	<50
103		Jiangsu	Nanjing	6	1000	2015.12	<5	<35	<50

*No data.

References

[1] I. Morris, The Measure of Civilization—How Social Development Decides the Fate of Nations, Princeton University Press, 2013.
[2] Energy and Environment—The European Approach, Klaus Hein, 04/2010.
[3] United Nations, Department of Economic and Social Affairs, Population Division. World Population Prospects: The 2017 Revision, DVD Edition, 2017.
[4] BP plc, BP Statistical Review of World Energy 2001, BP p.l.c., London, 2001. in review.
[5] BP plc, BP Statistical Review of World Energy 2018, BP p.l.c., London, 2018. in review.
[6] https://en.wikipedia.org/wiki/List_of_countries_by_energy_consumption_per_capita.
[7] https://data.worldbank.org/indicator/EG.USE.PCAP.KG.OE.
[8] V. Sivaram, India's energy policy may determine world fate, Sci. Am. (Chinese Version) 138 (2017) 68–73.
[9] U.S. Energy Information Administration, Monthly Energy Review—Table 2.1.
[10] http://data.stats.gov.cn/easyquery.htm?cn=C01.
[11] OECD/IEA, India Energy Outlook-World Energy Outlook Special Report, 2015.
[12] BP plc, BP Statistical Review of World Energy 2016, BP p.l.c., London, 2016. in review.
[13] U.S. Energy Information Administration, International Energy Outlook 2016, 2016.
[14] BP plc, BP Statistical Review of World Energy 2017, BP p.l.c., London, 2017. in review.
[15] IEA, World Energy Balances (2015 edition), www.iea.org, 2015.
[16] IEA, Energy and Air Pollution: World Energy Outlook Special Report 2016, 2016.
[17] China Electricity Council, Annual Development Report of China Electric Power Industry 2015, 2016.
[18] C. Li, C. Mclinden, V. Fioletov, N. Krotkov, S. Carn, J. Joiner, D. Streets, H. He, X.R. Ren, Z.Q. Li, R.R. Dickeeson, India is overtaking China as the world's largest emitter of anthropogenic sulfur dioxide, Sci. Rep. 7 (2017) 14304.
[19] H.Z. Tian, L. Lu, J.M. Hao, J.J. Gao, K. Cheng, K.Y. Liu, P.P. Qiu, C.Y. Zhu, A review of key hazardous trace elements in Chinese coals: abundance, occurrence, behavior during coal combustion and their environmental impacts, Energy Fuel 27 (2) (2013) 601–614.
[20] M.L. Hui, L. Zhang, Z.G. Wang, S.X. Wang, The mercury mass flow and emissions of coal-fired power plants in China, China Environ. Sci. 35 (8) (2015) 2241–2250.
[21] Health Effects Institute, State of Global Air 2018: A Special Report on Global Exposure to Air Pollution and Its Disease Burden, Health Effects Institute, Boston, MA, 2018.
[22] Ministry of ecology and environment of the People's Republic of China, http://www.mee.gov.cn/hjzl/zghjzkgb/lnzghjzkgb/.
[23] S.M. Wang, J.Z. Liu. Investigation of near-zero air pollutant emission characteristics from coal-fired power plants. Proc. CSEE, 36(22): 6140-6148.
[24] S.M. Wang, C. Song, Y.B. Chen, P. Sun, Technology research and engineering applications of near-zero air pollutant emission coal-fired power plants, Res. Environ. Sci. 28 (4) (2015) 487–494.
[25] J.Z. ' .C.H.Z. ' .Y.X.Z. ' .G.C. Wu, S.Q. Zhu, W. Meng, X. Gao, K.F. Cen, Experimental investigation of ultra-low pollutants emission characteristics from a 1000MW coal-fired power plan, Proc. CSEE 36 (5) (2016) 1310–1314.
[26] China Environmental Protection Industry Association, Development Report of Desulfurization and Denitrification Industry 2016, 2017.
[27] China Electricity Council, China Coal and Electricity Clean Development Report, China Electric Power Press, 2017.

CHAPTER TWO

Coal-fired power plants emission standards

Tao Wang
School of Energy, Power and Mechanical Engineering, North China Electric Power University, Beijing, China

As mentioned in the Chapter 1, the ultra-low emission was a pollutant emission reduction initiative proposed by power generation enterprises, which finally became the national general requirement. Although it has not yet become the national emission standard, it still has made a significant effect on reducing air pollutants in China.

In fact, with the development of industrial society, the pollutant emission standards have always played an important role. The requirements for pollutant emission in different countries are different during historical stages, which are subject to the level of knowledge and environmental acceptable capacity.

Even though the emission standards have been introduced throughout the world, there are significant differences in their stringency from country to country. China, Korea, Japan, the United States, and European Union currently have the most stringent emission standards in the world.

The pollution emission standards will be discussed in this chapter for the typical coal-fired power generation countries. Many countries or regions have formulated the national (regional), local, and source-level emission standards related to different emission sources in different regions; among them, the national industry emission standards are the most important. National industry standards are a kind of technology-based emission standard whose establishment procedures and methods are the most stringent; they are also the minimum standards that need to be implemented nationwide. Therefore, they are not directly targeted at achieving air quality standards, but at achieving continuous and stable emission levels under the premise of cost effectiveness, as they are the guiding principle for the formulation of emission standards at other levels.

Advances in Ultra-low Emission Control Technologies for Coal-Fired Power Plants
https://doi.org/10.1016/B978-0-08-102418-8.00002-4

2.1 Emission standards for coal-fired power plants in China

With the development of economy and society, China began to focus on the environmental protection issues in the 1970s. Table 2.1 lists the historical evolution of environmental protection regulations and related power emission standards in China.

In 1973, China promulgated the *Trial Standard of Industrial "Three Wastes" Discharge* (GBJ4-73), which set the limit requirements of the emission of air pollutants from thermal power plants for the first time. In 1991, Ministry of Environmental Protection of China issued the *Emission standards of air pollutants for Coal-fired power plants* (GB13223-1991), and subsequently revised the Standards in 1996, 2003 and 2011. Currently, the *Emission standard of air pollutants for thermal power plants* (GB13223-2011) is being implemented, as shown in Table 2.2. The control of PM and SO_2 was carried out by GB13223-1991, NO_x was added in GB13223-1996, and GB13223-2011 set the emission limits for mercury and its chemical compounds. In general, more and more pollutant emission control items are being implemented with the requirements becoming more and more strict. The standards prior to GB13223-2003 set the different emission limits for coals with different ash content. In GB13223-2003 and subsequent standards, they required that no matter what type of coal is used, the emission must be lower than the corresponding limit without considering the difference in ash content any longer.

SO_2 concentration in the area around the stack of coal-fired power plants was calculated through Gaussian Diffusion Model and compared with the set emission limits in GB13223-1991, and GB13223-1996 considered wind

Table 2.1 Historical evolution of environmental regulations and power plant emission standards in China

Time	Emission standard of power plant
1973	Trial Standard of Industrial "Three Wastes" Discharge (GBJ4-73)
1991	Emission standards of air pollutants for Coal-fired power plants (GB13223-1991)
1996	Emission standard of air pollutants for thermal power plants (GB13223-1996)
2003	Emission standard of air pollutants for thermal power plants (GB13223-2003)
2011	Emission standard of air pollutants for thermal power plants (GB13223-2011)

Note: In this table, GB means national standards.

Table 2.2 Standard limit for emission of air pollutants from coal-fired power plants (GB13223-2011) (mg/m^3)

	Implementation time	PM	SO_2	NO_x	Mercury
New boiler	2012.1.1	30	100 200[a]	100 200[b]	0.03[c]
Existing boiler	2014.7.1		200 400[a]		
Key areas	2012.1.1	20	50	100	0.03

[a]The coal-fired power boilers located in Guangxi Zhuang Autonomous Region, Chongqing Municipality, Sichuan Province, and Guizhou Province implement this standard limit;
[b]The coal-fired power boiler adopting W-type flame furnace, the existing coal-fired power boiler with circulating fluidized bed, and the coal-fired power boiler built and put into operation before December 31, 2003, or approved by the environmental impact statement of the construction project implement this standard limit;
[c]Since January 1, 2015, the coal-fired power boilers implement the mercury and its combined pollutant discharge limits specified in this table.

velocity and other conditions, which specified the maximum allowable SO_2 emissions from thermal power plants. However, these two SO_2 emissions monitoring methods are difficult to be used in practical applications, which was not conducive to environmental monitoring implementation. In GB13223-1996, coal-fired power plants were divided into 3 classes in chronological order, stipulated the maximum allowable emission concentration of SO_2 and NO_x from power plants, which were approved and constructed after 1997. However, GB13223-2003 standard required all coal-fired power plants to implement the corresponding PM, SO_2, and NO_x emission limits. The emission limits were further reduced in GB13223-2011 standard, the special emission limits for coal-fired boilers in key areas are 20 mg/m^3 of PM, 50 mg/m^3 of SO_2, and 100 mg/m^3 of NO_x.

2.2 Emission standards for coal-fired power plants in the United States

In June 1979, the United States proposed performance standards for the new stationary pollution sources of power plant steam-generating units, which was revised to the standards of performance for coal-fired steam generating units in September 1998. Since 2006, the *Environmental Protection Act* has been gradually improved and revised to the national emission standards for hazardous air pollutants from coal-fired power plants, industrial commercial organizations, and small industrial commercial organizations, as shown in Table 2.3. The latest version is a revised version from 2015; as

Table 2.3 Historical evolution of U.S environmental regulations and power emission standards

Time	Environmental regulations
1979.6	New Stationary Sources Performance Standards; Electric Utility Steam-Generating Units
1998.9	Revision of Standards of Performance for Nitrogen Oxide Emissions From New Fossil-Fuel Fired Steam-Generating Units; Revisions to Reporting requirements for Standards of Performance for New Fossil-Fuel Fired Steam-Generating Units
2001.4	Standards of Performance for Electric Utility Steam-Generating Units for Which Construction is Commenced After September 18, 1978; Standards of Performance for Industrial-Commercial-Institutional Steam-Generating Units
2006.2	Standards of Performance for Electric Utility Steam Generating Units, Industrial–Commercial–Institutional Steam Generating Units, and Small Industrial–Commercial–Institutional Steam Generating Units; Final Rule
2007.6	Amendments to New Source Performance Standards (NSPS) for Electric Utility Steam-Generating Units and Industrial-Commercial-Institutional Steam-Generating Units; Final Rule
2011.1	Standards of Performance for Fossil-Fuel-Fired, Electric Utility, Industrial-Commercial-Institutional, and Small Industrial-Commercial-Institutional Steam-Generating Units
2012.2	National Emission Standards for Hazardous Air Pollutants From Coal- and Oil-Fired Electric Utility Steam-Generating Units and Standards of Performance for Fossil-Fuel-Fired Electric Utility, Industrial-Commercial-Institutional, and Small Industrial-Commercial-Institutional Steam-Generating Units; Final Rule
2013.4	Reconsideration of Certain New Source and Startup/Shutdown Issues: National Emission Standards for Hazardous Air Pollutants From Coaland Oil-Fired Electric Utility Steam-Generating Units and Standards of Performance for Fossil-Fuel-Fired Electric Utility, Industrial-Commercial-Institutional, and Small industrial-Commercial-Institutional Steam-Generating Units
2015.2	National Emission Standards for Hazardous Air Pollutants From Coaland Oil-Fired Electric Utility Steam-Generating Units and Standards of Performance for Fossil-Fuel-Fired Electric Utility, Industrial-Commercial Institutional, and Small Industrial Commercial-Institutional Steam-Generating Units; Revisions

shown in Tables 2.4 and 2.5, they are the standard emission requirements for new upgraded and existing units.

The National Emission Standards for thermal power plants in the United States consisted of two parts, one part was the *New Source Performance*

Table 2.4 Emission limits for new or reconstructed EGUs of US

Type of unit	Pollutants	Emission limits	Monitoring method
Coal-fired unit not low-rank virgin coal	Filterable PM	9.0E-2 lb/MWh	Collect a minimum of 4 dscm per run
	Total non-Hg HAP metals	6.0E-2 lb/GWh	Collect a minimum of 4 dscm per run
	Individual HAP metals: Antimony (Sb) Arsenic (As) Beryllium (Be) Cadmium (Cd) Chromium (Cr) Cobalt (Co) Lead (Pb) Manganese (Mn) Nickel (Ni) Selenium (Se)	8.0E-3 lb/GWh 3.0E-3 lb/GWh 6.0E-4 lb/GWh 4.0E-4 lb/GWh 7.0E-3 lb/GWh 2.0E-3 lb/GWh 2.0E-2 lb/GWh 4.0E-3 lb/GWh 4.0E-2 lb/GWh 5.0E-2 lb/GWh	Collect a minimum of 3 dscm per run
	HCl	1.0E-2 lb/MWh	For Method26A, collect a minimum of 3 dscm per run. For ASTM D6348-032 or Method 320, sample for a minimum of 1 h
	SO_2	1.0 lb/MWh	SO_2 CEMS
	Mercury (Hg)	3.0E-3 lb/GWh	Hg CEMS or sorbent trap-Monitoring system only
Coal-fired units low-rank virgin coal	Filterable PM	9.0E-2 lb/MWh	Collect a minimum of 4 dscm per run
	Total non-Hg HAP metals	6.0E-2 lb/GWh	Collect a minimum of 4 dscm per run
	Individual HAP metals: Antimony (Sb) Arsenic (As) Beryllium (Be) Cadmium (Cd) Chromium (Cr) Cobalt (Co) Lead (Pb) Manganese (Mn) Nickel (Ni) Selenium (Se)	8.0E-3 lb/GWh 3.0E-3 lb/GWh 6.0E-4 lb/GWh 4.0E-4 lb/GWh 7.0E-3 lb/GWh 2.0E-3 lb/GWh 2.0E-2 lb/GWh 4.0E-3 lb/GWh 4.0E-2 lb/GWh 5.0E-2 lb/GWh	Collect a minimum of 3 dscm per run

Continued

Table 2.4 Emission limits for new or reconstructed EGUs of US—cont'd

Type of unit	Pollutants	Emission limits	Monitoring method
	HCl	1.0E-2 lb/MWh	For Method 26A, collect a minimum of 3 dscm per run. For ASTM D6348-032 or Method 320, sample for a minimum of 1 h
	SO_2	1.0 lb/MWh	SO_2 CEMS
	Mercury (Hg)	4.0E-2 lb/GWh	Hg CEMS or sorbent trap-monitoring system only

Table 2.5 Emission limits for existing EGUs of US

Type of unit	Pollutants	Emission limits	Monitoring method
Coal-fired unit not low-rank virgin coal	Filterable PM	3.0E-1 lb/MWh	Collect a minimum of 1 dscm per run
	Total non-Hg HAP metals	5.0E-1 lb/GWh	Collect a minimum of 1 dscm per run
	Individual HAP	8.0E-3 lb/GWh	Collect a minimum of 3 dscm per run
	metals:	2.0E-2 lb/GWh	
	Antimony (Sb)	2.0E-3lb/GWh	
	Arsenic (As)	3.0E-3 lb/GWh	
	Beryllium (Be)	3.0E-2 lb/GWh	
	Cadmium (Cd)	8.0E-3 lb/GWh	
	Chromium (Cr)	2.0E-2 lb/GWh	
	Cobalt (Co)	5.0E-2 lb/GWh	
	Lead (Pb)	4.0E-2 lb/GWh	
	Manganese (Mn)	6.0E-2 lb/GWh	
	Nickel (Ni)		
	Selenium (Se)		
	HCl	2.0E-2 lb/MWh	For Method 26A, collect a minimum of 0.75 dscm per run; for Method 26, collect a minimum of 120 liters per run. For ASTM D6348-033 or Method 320, sample for a minimum of 1 h

Table 2.5 Emission limits for existing EGUs of US—cont'd

Type of unit	Pollutants	Emission limits	Monitoring method
	SO_2	1.5E0 lb/MWh	SO_2 CEMS
	Mercury (Hg)	1.3E-2 lb/GWh	LEE Testing for 30 days with a sampling period consistent with Method 30B run or Hg CEMS or sorbent trap-monitoring system only
		1.1E-2 lb/GWh	LEE Testing for 90 days with a sampling period consistent with Method 30B run or Hg CEMS or sorbent trap-monitoring system only
Coal-fired units low-rank virgin coal	PM	3.0E-1 lb/MWh	Collect a minimum of 1 dscm per run
	Total non-Hg HAP metals	5.0E-1 lb/GWh	Collect a minimum of 1 dscm per run
	Individual HAP metals:		Collect a minimum of 3 dscm per run
	Antimony (Sb)	8.0E-3 lb/GWh	
	Arsenic (As)	2.0E-2 lb/GWh	
	Beryllium (Be)	2.0E-3 lb/GWh	
	Cadmium (Cd)	3.0E-3 lb/GWh	
	Chromium (Cr)	3.0E-2 lb/GWh	
	Cobalt (Co)	8.0E-3 lb/GWh	
	Lead (Pb)	2.0E-2 lb/GWh	
	Manganese (Mn)	5.0E-2 lb/GWh	
	Nickel (Ni)	4.0E-2 lb/GWh	
	Selenium (Se)	6.0E-2 lb/GWh	
	HCl	2.0E-2 lb/MWh	For Method 26A, collect a minimum of 0.75 dscm per run; for Method 26, collect a minimum of 120 liters per run. For ASTM D6348-03 or Method 320, sample for a minimum of 1 h
	SO_2	1.5E0 lb/MWh	SO_2 CEMS
	Mercury (Hg)	4.0E-2 lb/GWh	LEE Testing for 30 days with a sampling period consistent with Method 30B run or Hg CEMS or sorbent trap-monitoring system only

Note: LEE: low-emitting steam-generating units

Standards (NSPS) for conventional pollutant emission, and the other part was the *National Emission Standard for Hazardous Air Pollutant* (NESHAP). The core issue of the standards concerned the emission limits and their monitoring, recording, and reporting requirements. The content of NSPS standard in the United States was much more than that of China coal-fired power plants emission standards. The various emission limits were implemented in the United States according to the type of fuels and the age of construction of controlled units, and the corresponding monitoring, recording, and the reporting requirements were also more detailed and complete, which was conducive to the implementation of these regulations through the management of the discharge permit. The emission limits applied in the NSPS for thermal power plants in the United States included performance limits, removal rate limits, and opacity percentage limits, among them the most important was the performance limits, including the performance limits based on heat input and energy output. The performance limits were evaluated on the cycle of 30-day moving average scale. The most widely used was output-based performance limits, which was expressed in the form of emissions/power generation (lb/MWh) or emissions/generated steam heat (lb/ MMBtu heat output). The output-based limits took into account the cost effectiveness of emission reduction measures, such as improving combustion efficiency, increasing turbine efficiency, recovering useful heat, reducing parasitic losses associated with the operation of controlled power plants, encouraging controlled sources to reduce the use of fossil fuels, providing greater flexibility to the controlled sources for reducing source compliance cost.

NSPS and NESHAP for the coal-fired power plants in the United States are technology-based emission standards; they reflected the emission level that can be achieved at a certain technical level. However, China has not defined this basic requirement of "technology-based emission standards." The content of NSPS standard in the United States is much more detailed than that of the China coal-fired power plant emission standards, especially for the emission limit regulations, the United States implemented various emission limit regulations according to the fuels and the age of construction of controlled units, and the corresponding monitoring, recording, and the reporting requirements were also more detailed and complete, which was conducive to the implementation of these regulations through the management of discharge permit. NSPS for the US coal-fired power plant mainly adopts an output-based performance limit and adopts a long average period of 30-day moving average assessment method. However, the average value of 1 h is mainly assessed in the China National Standards, and if the requirements are too strict, it is difficult for power plants to ensure a small and stable fluctuation of emission concentration in a short time scale. Especially when the coal-fired unit in the stage of start-up or shutdown, it may be difficult to meet the requirements. The increase of environmental benefits by adopting more strict emission standards is not obvious, while it will have a negative impact on the unit running.

2.3 Emission standards for coal-fired power plants in Europe

For the specified emission limits given for the units with different capacities in Europe, the larger the unit capacity is, the stricter the emission requirements are. The most stringent emission limits for PM, SO_2, and NO_x of coal-fired units beyond 300 MW are 10 mg/m^3, 20 mg/m^3, 20 mg/m^3, respectively.

All permits for installations containing coal-fired power plants, which have been granted a permit before 7 January 2014, or the operators of which have submitted a complete application for a permit before that date, provided that such plants are put into operation no later than 7 January 2014, shall include conditions ensuring that emissions into air from these plants do not exceed the emission limit values set out in Table 2.6. Where a coal-fired power plant is extended, all permits for installations, which had been granted, stated that emissions from these plants should do not exceed the emission limit values set out in Table 2.7. The emission limit values for PM, SO_2, and NO_x set out in the permit for the coal-fired power plant applicable on 31 December 2015, pursuant in particular to the requirements of Directives 2001/80/EC and 2008/1/EC, shall at least be maintained.

Table 2.6 Emission limits for EU coal-fired power plants put into operation before 7 January 2014 (mg/Nm^3)

Capacity (MW)	SO_2	NO_x	PM
50–100	400	300	30
100–300	250	200 450 in case of pulverized lignite combustion	25
>300	200	200	20

Table 2.7 Emission limits for EU coal-fired power plants put into operation after 1 January 2016 (mg/Nm^3)

Capacity (MW)	SO_2	NO_x	PM
50–100	400	300 400 in case of pulverized lignite combustion	20
100–300	200	200	20
>300	150 200 in case of circulating or pressurized fluidized-bed combustion	150 200 in case of pulverized lignite combustion	10

2.4 Emission standards for coal-fired power plants in India

India's environmental law originated from the Indian Constitution of 1949. This Law had a power allocation for pollution prevention and ecological protection legislation. Of these, the Federal shared the legislative power for forests, wildlife protection, small ports, and factories. This Law was used throughout the postcolonial period to solve sporadic environmental problems.

After India began its industrialization, the Constitution was unable to cope with modern environmental problems. Therefore, India started its environmental legislation in the modern sense at the end of 1970. Prior to the Bhopal pesticide spill in 1984, India's major environmental legislation included *Wildlife Protection Act* in 1972, *Water Pollution Prevention Act* in 1974, *Forest Protection Act* in 1980, and *Air Pollution Prevention and Control Act* in 1981.

In May 1986, the Federal Parliament of India passed the first comprehensive environmental legislation *Environmental Protection Act*, which came into force on November 19, 1986. Although the legislative purpose stated in the preamble of the Act was to implement the 1972 *United Nations Declaration on the Human Environment*, the essence of the Act was due to the tremendous pressure imposed on India to enact the comprehensive environmental laws to manage the toxic and dangerous substances after the Bhopal Incident.

So far, India's environmental legislation has initially taken shape of a modern environmental legal system. In addition to the technical matters such as the scope of the adapted territory, the method and time of entry into force, and the legal status of the executive, the law covers the scope of

Table 2.8 Emission limits for coal-fired power plant of India (mg/Nm^3)

Units installed before 31st December, 2003[a]	
PM	**100**
SO_2	600 (Units Smaller than 500 MW capacity units) 200 (for units having capacity of 500 MW and above)
NO_x	600
Mercury (Hg)	0.03 (for units having capacity of 500 MW and above)
Units installed after 1st January, 2003, up to 31st December, 2016[a]	
PM	50
SO_2	600 (Units Smaller than 500 MW capacity units) 200 (for units having capacity of 500 MW and above)
NO_x	300
Mercury (Hg)	0.03
Units to be installed from 1st January, 2017[b]	
PM	30
SO_2	100
NO_x	100
Mercury (Hg)	0.03

[a]Units shall meet the limits within two years from date of publication of this notification.
[b]Includes all the units, which have been accorded environmental clearance and under construction.

application, the rights and obligations of the federal government, subsidiary administrative legislation, personal and corporate obligations and Legal liability, executive institutional arrangements and judicial enforcement.

The currently implemented standards provide the emission limits requirements for coal-fired power plant units at different time periods. The most stringent PM, SO_2, NO_x, and mercury emission limits for the newly built coal-fired units after 2017 are 30 mg/m^3, 100 mg/m^3, 100 mg/m^3 and 0.03 mg/m^3, respectively (Table 2.8).

2.5 Comparison of emission standards for coal-fired power plants in typical countries

In Table 2.9, these are the comparison of concentration limits of air pollutants such as PM, SO_2, and NO_x from coal-fired power plants in some countries in the world. The general trend is that the emission concentration of pollutants in new power plants is becoming increasingly stricter, and the limits of developing countries are still relatively high. China, South Korea, the European Union, the United States, and other countries and regions

Table 2.9 Comparison of emission limits of air pollutants in newly built large coal-fired power plants between typical countries and regions in the World [1]

Region	Policy	SO_2		NO_x		PM	
		Existing	New	Existing	New	Existing	New
China[a]	Emission standard of air pollutants for thermal power plants , 2011	100–400	50	100–200	100	30	20
European Union	Industrial Emissions Directive , 2010	200–400	150–400	200–450	150–400	20–30	10–20
United States[b]	New Source Performance Standards , 2015	160–640	160	117–640	117	23	23
India	Environment (Protection) Amendment Rules, 2015	200–600	100	300–600	100	50–100	30
Indonesia	MOE decree no. 21 2008	750	750	850	750	150	100
Japan[c]	Air Pollution Control Law 1993	–	–	123–513	123–513	30–100	30–100
Mexico[d]	Mexican Official Standard NOM-085-ECOL-1994 (in PPMV for SO_2 and NO_x)	550–2200	30–2200	110–375	25–375	60–450	60–450
Philippines	National Emission Standards for Particulate Matter for Stationary Sources	1000–1500	200–700	1000–1500	500–1000	150–200	150–200
South Africa	The Minimum Emissions Standards are published by the government	3500	500	1100	750	100	50
Korea	Special Measures for Metropolitan Air Quality Improvement , 2002	286	229	308	164	40	20–30
Thailand	Royal Thai Government Gazette	700–1300	180–360	400	200	80–320	80
Vietnam	Industrial emission standards for dust and inorganic substances, 2005	1500	500	1000	650–1000	400	200

[a] "New" Special emission limits for air pollutants applicable to key areas adopted in China.
[b] US emission limits were converted from lb/MBtu to mg/m^3 assuming an F-factor of 1800 standard cubic feet of CO_2 and a CO_2 content of 12% in the flue gas.
[c] Japan Air Pollution Control Law (APCL) specifies emission limits for SO_x, NO_x, and PM that differ depending on the scale of facilities, technologies, and regions. Prior to the construction of new plants in Japan, local authorities and power generation companies usually arrive at bilateral pollution prevention agreements more stringent than those mandated in the APCL, 1993.
[d] For Mexico, SO_2 and NO_X are expressed in parts per million by volume (PPMV).
Notes: "Existing" refers to the emission limit for currently operating power plants. "New" refers to the limit for planned or proposed plants.

have strict requirements for the emission of air pollutants from coal-fired power plants. In these countries, China has the highest emission requirements regarding the conventional pollutants such as PM, SO_2, and NO_x, but the US emission requirements are more detailed, and more specific regulations are imposed on other pollutant emissions.

2.6 Ultra-low emission

As a major coal consumer, China coal-fired power plants consumed a large amount of coal, resulting in a huge amount of pollutant emissions. In order to protect the environment, China had to enhance the requirement of pollutants emission. Since 2013, China power generation enterprises have initiated and promoted "Near-zero emission" and "Ultra-low emission" for the coal-fired power plants [2–5].

The "Near-zero emission" refers to the emission concentration of PM, SO_2, and NO_x not higher than 5 mg/m^3, 35 mg/m^3, 50 mg/m under the condition of 6% of reference oxygen content, and the "Ultra-low emission" refers to the emission concentration of PM, SO_2, and NO_x not higher than 10 mg/m^3, 35 mg/m^3, 50 mg/m^3 under the condition of 6% of reference oxygen content, respectively. In view of the emission values, it is possible to achieve the requirements of the gas turbines emission standards specified in the "Emission standard of air pollutants for thermal power plants" GB13223-2011. From 2013 to 2015, Tianjin Dagang Power Plant, Hebei Sanhe Power Plant, Zhejiang Zhoushan Power Plant, and other power plants have achieved these emission targets [6].

In March 2015, the Premier's Government Work Report of China at the Third Session of the Twelfth National People's Congress of the State Council called for the further implementation of the air pollution prevention and control action plan. By 2020, the ultra-low emissions of coal-fired units will be fully implemented, and the coal consumption and pollution emissions of coal-fired power plants will be greatly reduced. Although "ultra-low emission" is not yet a national standard, it has played a role similar to the national standards.

References

[1] IEA, Energy and Air Pollution, World Energy Outlook Special Report, 2016.
[2] S.M. Wang, C. Song, Y.B. Chen, P. Sun, Technology research and engineering applications of near-zero air pollutant emission coal-fired power plants, Res. Environ. Sci. 28 (4) (2015) 487–494.

[3] J. Zhang, C.H. Zheng, Y.X. Zhang, G.C. Wu, S.Q. Zhu, W. Meng, X. Gao, K.F. Cen, Experimental investigation of ultra-low pollutants emission characteristics from a 1000MW coal-fired power plan, Proc. CSEE 36 (5) (2016) 1310–1314.
[4] Z.F. Sui, Y.S. Zhang, Y. Peng, P. Norris, Y. Cao, W.P. Pan, Fine particulate matter emission and size distribution characteristics in an ultra-low emission power plant, Fuel 185 (2016) 863–871.
[5] L.Q. Wang, F.H. Zhu, X.Y. Zhao, Potential capabilities of pollutant reduction and environmental benefits from ultra-low emissions of coal-fired power plants, Electric Power 47 (11) (2014) 150–154.
[6] S.M. Wang, Research on Coal-Fired Power Plants Near-Zero Emission Technologies and Applications, North China Electric Power University, 2017.

CHAPTER THREE

Key technologies for ultra-low emissions from coal-fired power plants

Carlos E. Romero, Xingchao Wang
Energy Research Center, Lehigh University, Bethlehem, PA, United States

Coal remains an abundant energy source in the world. Worldwide coal consumption will remain approximately the same over the next 20 years (2020–2040), with decreasing consumption in China and the U.S. offsetting increases in India [1]. Raw coal contains elements such as carbon, sulfur, nitrogen, and other elements. Due to the coal combustion process, coal-fired power plants are one of the main sources of sulfur dioxides (SO_2), nitrogen oxides (NO_x), and dust (also known as particulate matter, PM) emissions, which have serious impacts on the environment and human health. In order to limit the emissions of SO_2, NO_x, and PM, emission control technologies have been developed and deployed in the coal-fired power generation industry. A typical coal-fired power plant with conventional emissions control technologies is illustrated in Fig. 3.1. These techniques include low-NO_x burners (LNBs) and selective catalytic reduction (SCR) for NO_x control, electrostatic precipitator (ESP), or fabric filter (FF) for PM control, and flue gas desulfurization (FGD) for SO_2 control.

After 2010s, new regulations in China, the US, and the EU urgently required coal-fired power plants to achieve new and more stringent lower emissions limits. To achieve these goals, more advanced and efficient emissions control technologies are necessary to be investigated and proven. In this chapter, key technologies developed for coal-fired power plant ultra-low emissions are introduced and discussed.

3.1 Air pollutant emissions control technologies

Emissions of SO_2, NO_x, and PM caused by coal combustion processes are the main air pollution sources in the world. SO_2 originates from the sulfur content in the coal and is formed during coal burning. The majority of

Advances in Ultra-low Emission Control Technologies for Coal-Fired Power Plants
https://doi.org/10.1016/B978-0-08-102418-8.00003-6

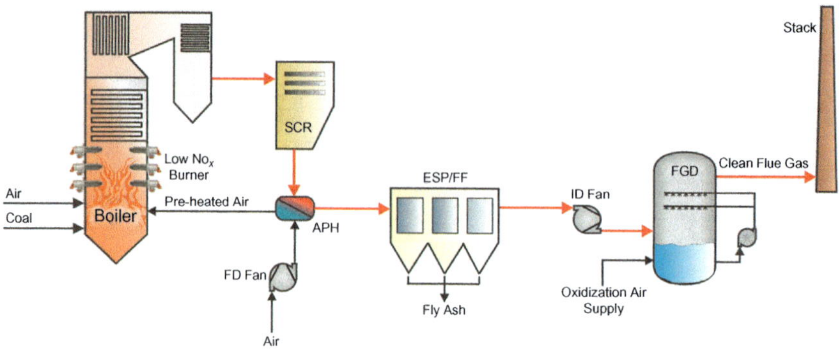

Fig. 3.1 Typical coal-fired power plant flue gas emissions control train.

the sulfur content in coal will yield SO_2. However, some of the SO_2 is further oxidized to SO_3 in the boiler and SCR. This can be represented by the following equations:

$$S(\text{in coal}) + O_2 \rightarrow SO_2 \tag{3.1}$$

$$2SO_2 + O_2 \rightarrow 2SO_3 \tag{3.2}$$

Sulfur content in coal ranges from 0.5% to more than 5% depending on the type of coal [2]. Current precombustion cleaning technology to clean hard coal can remove approximately 50% of the inorganic sulfur, but it leaves uncleaned the organic fraction of sulfur in coal. Therefore, coal-fired power plants rely on flue gas desulfurization as the primary means for the control of SO_x (SO_2, SO_3) emissions. There are different types of flue gas desulfurization systems for SO_x emissions control in coal-fired power plants.

NO_x refers to the emissions of nitric oxide (NO), nitrogen dioxide (NO_2), and other trace amount of nitrogen species formed during combustion process. The combustion process generates NO_x because of high combustion temperatures and availability of oxygen and nitrogen, both from the air and coal. N_2 and O_2 combine to form NO and NO_2 at temperatures up to 1600°C, which is termed as thermal NO_x formation. NO_x formed due to the nitrogen content in the fuel is called as fuel NO_x. The equations describing these processes are:

$$N_2 + O \rightarrow NO + N \tag{3.3}$$

$$N + O_2 \rightarrow NO + O \tag{3.4}$$

$$2NO + O_2 \rightarrow 2NO_2 \tag{3.5}$$

The major denitrification technologies in coal-fired power plants are low-NO_x burners, SCR, selective noncatalytic reduction (SNCR), and a combination of these technologies.

PM is typically classified as $PM_{2.5}$ and PM_{10}. $PM_{2.5}$ refers to particulate matter with a diameter of 2.5 μm or less and PM_{10} is particulate matter with a diameter of 10 μm or less, including $PM_{2.5}$. $PM_{2.5}$ is considered to have more health impacts than the relatively large particles. Exposure to fine particles can affect lung function and worsen respiratory medical conditions. People with breathing and heart problems, children, and elderly people are particularly sensitive to $PM_{2.5}$. The majorities of large economies have enforced air quality standards for PM. PM from coal-fired power plants is mainly contributed by coal fly ash (referred to as primary particles) and some by sulfates and nitrates (referred to as secondary particles). Particularly, the coal fly ash directly emitted to the atmosphere makes up the primary particles; gaseous SO_2/NO_x emitted by coal-fired power plants reacts with O_2 and H_2O droplets in the air to form the secondary particles. In order to control the PM concentration in ambient air, fly ash from coal-fired power plants has to be reduced, which requires installation of PM control facilities. Additionally, the synergy among the different emissions control system can also contribute to effectively reduce all types of flue gas emissions from coal-fired power plants.

3.1.1 PM control technologies

The PM removal technologies discussed in this section are mainly postcombustion methods. Postcombustion PM removal equipment removes and collects PM from the flue gas downstream of the boiler (Fig. 3.1). Conventional PM removal technologies are Electrostatic Precipitators (ESPs) and Fabric Filters (FFs). ESP and FF have proven to be the most suitable PM control technology for coal-fired power plants. The operating principle of ESPs and FFs is introduced in this section.

3.1.1.1 Electrostatic precipitators (ESPs)

Operating principle of ESPs

There are three fundamental steps in the ESP particulate removal process: particle charging, particle collection, and removal of the collected particle. In this section, the principles of ESP operation are introduced.

An ESP (Fig. 3.2) establishes an electromagnetic field by using a direct-current voltage across a pair of electrodes, one is a discharge electrode and the other one is a collection electrode. The power supply for the ESP is

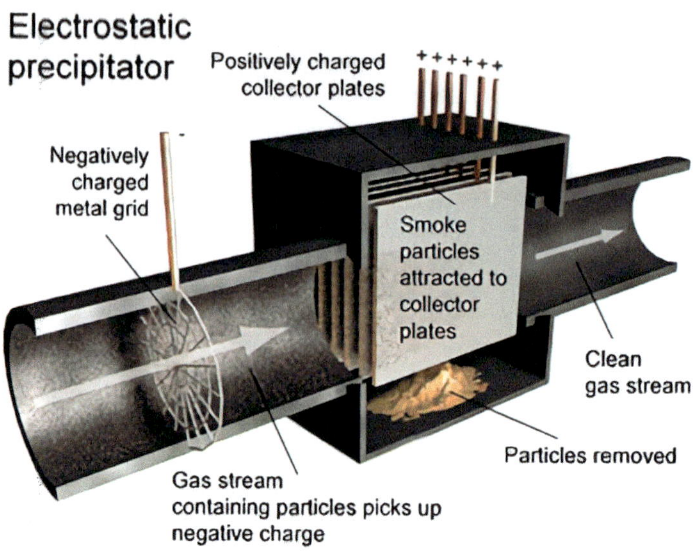

Fig. 3.2 Side view of a typical dry ESP. *(From https://www.qld.gov.au/__data/assets/image/0026/66824/electrostatic.jpg.)*

converted from the source AC voltage (typically 220–480 V), to a pulsating DC voltage (20,000–100,000 V). The gas between the electrodes is ionized by applying this high voltage, which is known as a *corona* and the ionized particles become part of an electrical circuit. The ionized particles surround the discharge electrode in a circular region, and this phenomenon is called *corona discharge*, which is responsible for charging the particles. Particles enter into the ESP with the flue gas and are negatively charged by the discharge electrode to become ions. Then, the ions migrate by electrical force and stick to the collection electrode. To operate continuously, the collected particles have to be removed periodically or continuously. In a dry ESP (DESP), the particles are periodically loosened and knocked down by rappers and fall into a collection hopper, this process is called *rapping*. In a Wet ESP (WESP), the accumulated particles are washed away by a water flow. Around the collection area, there are some space or necessary clearances with no electrical force at the top of the ESP and above the hoppers to support the electrodes. The flue gas may flow through these areas where the particles are not being collected, and this process is known as *sneakage*. This also places an upper limit on the removal efficiency of an ESP [3].

The operating performance of an ESP mainly depends on three aspects: (1) fly ash particle properties, (2) applied voltage to the ESP and (3) flue gas characteristics.

Particle Properties—The primary particle properties to affect ESP removal efficiency are particle size distribution and fly ash resistivity. The first important particle property is size distribution. Size distribution describes how many particles of a given size there are, which is important because ESP efficiency varies with particle size. Most of the large particles with diameter of >1 μm are negatively charged by capturing negatively charged gas ions generated with a corona. This process is referred to as *direct charging* or *field charging*. These ions then migrate toward the collection plates. Smaller particles are charged by these ions while they are passing by, which is known as *diffusion charging*. Diffusion charging is related to the random Brownian motion of the negatively charged ions. This random motion is associated with thermal effects. Diffusion charging mainly charges particles with a diameter less than 0.3 μm. For particles with an intermediate diameter of 0.3–1 μm, a combination of these two charging mechanisms occurs. However, particles with a diameter within the range from 0.3 to 0.6 μm are hard to be charged due to the transition between field charging and diffusion charging [4]. The second important parameter that affects ESP removal efficiency is fly ash resistivity. It stands for how well the particulate deposited on the ESP collection electrode conducts electricity. For fly ash with normal resistivity ($1 \times 10^7 - 2 \times 10^{10}\ \Omega \cdot \text{cm}$), collection is deemed stable. Most of the fly ash handled by ESP is in this electrical resistivity range. Low-resistivity fly ash ($<1 \times 10^7 \Omega \cdot \text{cm}$) cannot hold sufficient charge to be captured by the collection electrode; thus, they are re-entrained in the collection area, and the ESP removal efficiency is significantly reduced. This process is called *re-entrainment*. For fly ash with high resistivity ($>2 \times 10^{10}\ \Omega \cdot \text{cm}$), the portion of high-resistivity particles increase will result in decreasing removal efficiency due to low migration velocity. Additionally, the applied voltage and the discharge current are reduced to cause a flow of electrons back to the discharge electrode, which is referred to as *sparking* (also known as flashover, spark-over). Sparking reduces PM removal efficiency significantly because no charging occurs when sparking happens. Subsequently, there is no force to hold the particles on the collection plates to lead to re-entrainment. However, when the voltage drop through the ash layer is large enough, there is an electrical break down that results in generating positive ions, which create a corona current as shown in Fig. 3.3. This corona current moves toward the discharge electrode, which is the undesirable direction. This phenomenon is called *back corona* (Fig. 3.3). Therefore, the resistivity of fly ash being collected is an important factor in designing an ESP.

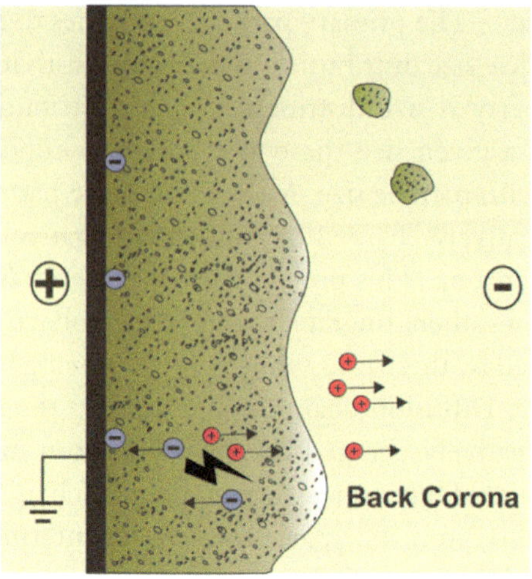

Fig. 3.3 Back corona in ESPs.

Voltage applied to the ESP—In general, an ESP is desired to operate at a voltage as high as possible. The reason why the electrodes are charged negatively is that a negative corona supports a higher voltage than a positive corona before sparking occurs. However, high voltages can also cause sparking, which result in an instantaneous termination of the electric field. As the voltage increases, the particle charging and the sparking increases. An ESP working at optimal conditions has an average sparking rate of 50 to 100 sparks per minute [4]. When the sparking rate exceeds this range, it generally leads to a decrease in collection efficiency. The lowest acceptable applied voltage is the voltage required to form a corona, the electrical discharge that produces ions for charging particles.

Flue gas characteristics—Characteristics that include the flue gas flow rate, flue gas temperature, and flue gas composition have an influence on fly ash emissions level or ESP collection efficiency. The flue gas flow velocity through an ESP is an indicator of residence time. ESP collection efficiency is a function of residence time. For example, the particle concentration in the ESP outlet may double with an increase of 20% in flue gas flow rate. The higher collection efficiency is obtained with the longer residence time. However, if the flue gas flow velocity is too large, particles in an ESP are potentially re-entrained. Typically, design criteria for velocity are about 1.5 m/s for plate-wire ESPs and 1 m/s for flat-plate ESPs, which are introduced subsequently. The flue gas temperature can also have a strong effect on the resistivity of the particles in two ways. Firstly, the gas is denser, which increases the specific collection area. Secondly, moisture is condensed on the particles due to lower flue gas temperature. Finally, the flue gas composition, including ash concentration, sulfur content, moisture content, and ash composition, can also impact ESP collection efficiency. Higher particle concentrations need a higher voltage to be applied to the ESP to maintain the same level of PM emissions. In terms of sulfur content, H_2SO_4 would form and lay down on the particles creating a conductive layer that would reduce fly ash resistivity. Low sulfur flue gas content has a higher resistivity. Similar to the sulfur content, a high moisture content in the flue gas leads to low fly ash resistivity. Ref. [5] concludes that low sodium, iron, lithium, and high calcium content cause high fly ash resistivity. It is possible to treat the fly as for resistivity conditioning, this is typically considered to compensate for changes to design conditions and other factors influencing ESP performance [5].

To estimate the performance of ESPs, the theoretical efficiency of ESP can be obtained by the Deutsch-Anderson equation:

$$\eta = 1 - e^{-w\frac{A}{Q}}, \tag{3.6}$$

where:

η = Fractional collection efficiency
w = Terminal drift velocity in m/s
A = Total collection area in m^2
Q = Volumetric air flow rate in m^3/s

This equation has been widely used to estimate the collection efficiency of ESPs. However, three main assumptions are made for calculations to establish preliminary ESP designs: (1) the fly ash re-entrainment is neglected during rapping; (2) it assumes that the particle size and, consequently, the migration velocity of fly ash is the same for all particles in the flue gas. *Migration velocity* is the average velocity of an individual charged particle migrating to the collection electrode (the migration velocity is not same for particles with different sizes); (3) uniform flue gas flow rate is assumed everywhere in the ESP, sneakage is neglected. Therefore, the Deutsch-Anderson equation is only used for preliminary estimation of ESP efficiency. More accurate estimates of collection efficiency can be obtained using the Matts-Öhnfeldt equation, which is modified from the Deutsch-Anderson equation:

$$\eta = 1 - e^{-w_k\left(\frac{A}{Q}\right)^k}, \tag{3.7}$$

where:

η = Fractional collection efficiency

w_k = Average migration velocity in m/s

k = Constant, usually 0.4 to 0.6

A = Total collection area in m^2

Q = Volumetric air flow rate in m^3/s

Specific collecting area (SCA) is another parameter used to roughly estimate ESPs collection efficiency. SCA is defined as the total collection area divided by gas volume flow rate (A/Q) in units of s/m. SCA is also one of the most important factors in determining capital and some O&M costs (for example, maintenance and dust disposal costs) of the ESP since it reflects the size of the unit.

Classification of ESPs

ESPs can be classified in several ways:

- By structure or design: plate-wire ESP, flat-plate ESP, tubular ESP

Plate-wire ESPs are widely used in industrial applications, including coal-fired power generation. In a plate-wire ESP, gas passes between parallel plates of sheet metal and high-voltage electrodes. The high-voltage electrodes are long wires hanging between the plates. They are also weighted and supported by rigid frames structure. The plate-wire ESP can have many channels to operate in parallel, which also can have a large height. As a consequence, the plate-wire ESP is suitable for handling large volumes of gases.

A flat-plate ESP consists of a series of parallel flat plates. Similar to plate-wire ESPs, some plates serve as high-voltage electrodes instead of wires in plate-wire ESPs. A flat-plate ESP increases the surface area for PM collection and also the electric field. Different from plate-wire and tubular ESPs, flat-plate ESPs can operate with either negative or positive polarity. Ozone generation is reduced when a positive polarity is used. In addition, corona cannot be generated by flat plats alone. Therefore, corona-generating electrodes are located ahead of and sometimes after the discharge plate zones. The corona-generating electrodes are independent corona wires or sharp pointed needles in the edges of the plates. A flat-plate ESP operates with little or no corona current flowing through the collected PM, except directly under the corona needles or wires, which makes the flat-plate ESP less affected by back corona than other types of ESPs. However, higher rapping losses happen in this design due to low electric force caused by a lack of current in the collected layer, which tends to remove the layer from the collecting surface. Flat-plate ESPs are widely used for PM with high resistivity but small mass median diameter, MMD (1–2 μm) (MMD is the diameter for which one-half of the particulate mass consists of smaller particles and one-half is larger).

A tubular ESP is basically a one-stage unit with a high-voltage electrode operated at one voltage running along the axis of a tube. Currently, only a small portion of ESPs in coal-fired power plants are tubular ESPs. They are particularly used where the particulate is either wet or sticky. There are no sneakage paths and tubular ESPs are usually cleaned with water; therefore, the re-entrainment losses are much lower than DESPs.

- By operation temperature: cold-side ESP, hot-side ESP

Cold-side ESPs are located after the air preheater (APH) and before the induced draft fan (ID fan). It typically operates within 120–180°C. Approximate 80% of the pulverized coal-fired plants are equipped with ESPs are with cold-side ESPs.

Hot-side ESPs are located upstream of the APH and operate in the 300–450°C range. Hot-side ESP is mainly used in the U.S and Japan and a 1990 study showed 150 hot-side ESPs were built in the U.S. during 1935 to 1990. However, the hot-side ESP has issues associated with the layer of high resistivity fly built up on the plate electrodes that needs sometimes to be sand blasted at month-long intervals. By 2008, almost all hot-side ESPs have been converted to run as cold-side ESPs [6].

- By particle removal method from collection surface: DESP, WESP

The main difference between DESP and WESP is the way to remove the collected particles from the collection surface, in the case of a WESP, using water as a cleaning method.

Advantages and disadvantages of ESPs

The advantages of using ESPs are: (1) high PM removal efficiency, (2) low O&M costs, (3) low pressure drop of the flue gas (0.1–0.3 kPa), (4) capable of treating large volumes of flue gas, (5) treatment time is negligible (0.1–10 s) and (6) cleaning is comparatively easy.

The disadvantages of using ESPs are: (1) high capital cost, (2) large space required due to equipment size, (3) ozone is produced during gas ionization,

(4) operating conditions such as particulate conductivity, sparking rate, flow velocity, and proper gas flow distribution must be maintained carefully.

3.1.1.2 Fabric filters (FFs)

Operating principle of FFs

FF (Fig. 3.4) has been typically used for PM control in smaller industrial processes. However, due to its effectiveness and higher collection efficiencies for fine particles, FFs are becoming more recently popular for use in coal-fired power plants. A FF can ensure >99% removal efficiency for fine particles and >99.5% for overall removal efficiency [7]. The principle of FF operation is filtration, which is efficient, reliable, and economic. In a FF, flue gas passes through a felted or woven fabric, and then the PM is collected on the fabric. Cylindrical bags are the most common type of fabric filter; therefore, FFs are typically referred to as baghouses. A bag house consists of number of tubular bags, which are vertically hanged. The collected fly ash then goes to hoppers to be stored and cleaned periodically.

Woven and felted fabrics are used to make bag filters. Woven filters are used with low-energy cleaning methods such as mechanical shaking and reverse air. Felted fabrics are usually used with higher energy-cleaning systems such as pulse jet cleaning. In selecting FF materials, some particulate and flue gas characteristics should be considered: (1) flue gas temperature, (2) flue gas flow rate, (3) flue gas composition, (4) particle

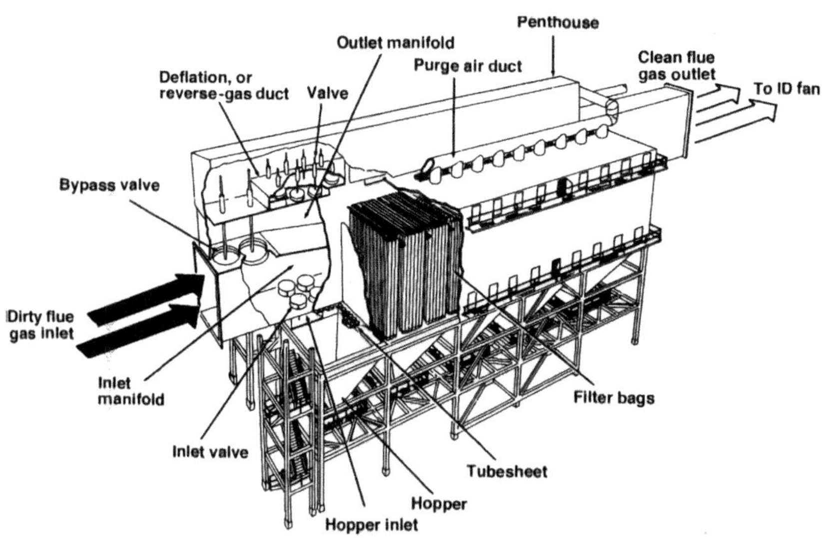

Fig. 3.4 Typical fabric filter. *(From G. Bruce, Clean Coal Engineering Technology, Elsevier Science & Technology, 2016.)*

size, distribution, and concentration. In addition, the fabric cost, abrasion resistance, chemical resistance, permeability, and tensile strength also need to be examined.

In a FF, the flue gas entering the inlet nozzle strikes a baffle plate, which causes larger particles to fall into hoppers due to gravity. The flue gas then flows upward into the tubes and outward through the fabric, leaving a layer of PM on the bags, as a *dust cake* (Fig. 3.4) (Bustard et al., 1988, cited in Bruce [8]). PM removal efficiency of a FF before and during the cake formation is low, but increases as the cake is formed, until a final efficiency exceeding 99% or even 99.9% is obtained. Once the cake is formed, it can be seen as part of the filtering media, which helps to boost the removal performance. The accumulated fly ash can be seen as filtering media. Generally, this cake is as important as the actual filter medium. The cake is usually formed within minutes or even seconds of operation. After a period of operating time, the accumulation of fly ash increases the air resistance of the filter and the bags have to be periodically cleaned. The dust cake will then be detached and fall into hoppers due to the cleaning process.

Fabric filters can be categorized by several ways, such as the type of cleaning (shaking, reverse air, pulse jet), direction of flue gas flow (from inside the bag toward the outside or reversely), location of the fan (ID or forced draft, FD, fan), or size (low, medium, or high gas flow quantity). Among all these categories, the type of cleaning is typically used for FF classification.

Operational considerations of FFs

A number of factors are considered for the operation of a bag house, including the gas to cloth (G/C) ratio, cleaning frequency, flue gas temperature, flue gas humidity, flue gas chemical characteristics, bag leakage, pressure drop, and continuous operation.

Gas to cloth (G/C) ratio—The *G/C ratio* is the amount of flue gas passing through a unit area of fabric in the baghouse. The factors that impact on the G/C ratio include the filter medium, fly ash size, fly ash density, fly ash concentration, and cleaning method. The G/C is also referred to as *face velocity* or *filter velocity*. The normal range of face velocity at which the gas is passed through the bags is 0.5–5 cm/s [9]. As the G/C ratio increases, the pressure drop also increases. Excessive G/C ratios lower particulate removal efficiencies and also lead to bags abrasion. Therefore, G/C ratio generally should be low to obtain high removal efficiencies. Low G/C ratios are also recommended for high concentration of particulate.

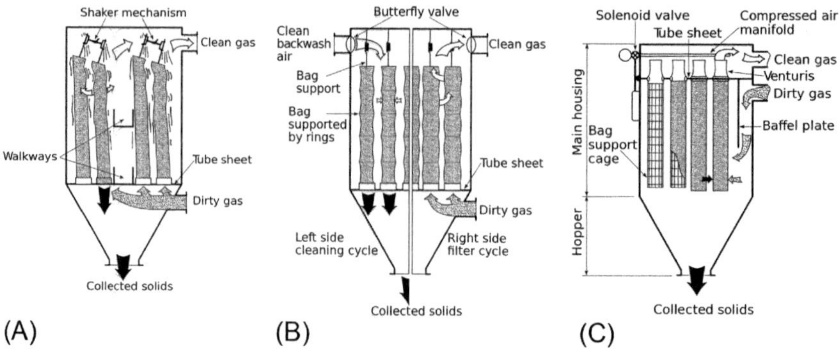

Fig. 3.5 FF cleaning method: (A) Mechanical shaking, (B) reverse air, (C) pulse jet [9]. *(From https://en.wikipedia.org/wiki/Baghouse.)*

Cleaning—At intervals, the bags get built up with a covering of dust particles that the gas can no longer pass through the bags. In this circumstance, the bags have to be cleaned by mechanical shaking or by a pulse jet reverse air flow. Cleaning methods for a FF are: (1) mechanical shaking, (2) reverse air flow, and (3) pulse jet (Fig. 3.5). The *Mechanical shaking* cleaning method is used with the flue gas flowing inside the bag toward the outside. During the shaking process, the pressure in the bags should not be positive. Compared with other cleaning methods, the shaking FF system tends to stress the bag filters more, which requires more durable fabric. Typically, the shaking method has a relatively low G/C Ratio, which results in requiring a larger space. The structure of *reverse air flow* cleaning is similar to mechanical shaking. The flue gas enters the baghouse and flows into the bags, leaving the fly ash inside the bags. To clean the bags, the flue gas flows through the compartments and then cleaning air is introduced to flow in the reverse direction. The deposited fly ash falls into hoppers. *Pulse jet* cleaning consists of directing a compressed air pulse to some of bags while the other bags can be still operating. The pulse jet–cleaning method is relatively new compared to other cleaning methods. During pulse jet cleaning, a short burst with duration of 0.03 to 1 s of high-pressure air is injected into the bags. The pulse creates a shock wave that moves from top to bottom of the bag. The wave flexes the fabric, pushing it away from the cage, and then snaps it back, dislodging the dust cake [10]. Pulse jet FFs have smaller baghouse sizes compared to baghouses with shaking and reverse air flow. The reasons are related to the no need to build extra compartments for offline cleaning and the higher G/C ratios. However, the larger pressure drop due to the higher G/C ratio leads to an increase of operation cost.

Flue gas temperature—FFs will not perform properly if the flue gas temperature exceeds the limitation of the fabric media. If the flue gas temperature is expected to fluctuate, a fiber material should be selected to sustain the upper temperature fluctuation. In addition, if the temperature in the baghouse falls below the acid dew point, this may need an additional heater to heat up the flue gas above the corresponding acid gas dew point.

Flue gas humidity—Maintaining baghouses at moisture-free conditions is important. Typically, many solids in the flue gas system are deliquescent, and if moisture is present, these particles tend to pick up moisture and dissolve slightly, which results in blinding of the filter. In such cases, it is extremely hard to remove this "mud" from the filter cloth without washing the filter.

Flue gas chemical characteristics—This is another problem associated with fabric filters. The possibility of chemical attack due to corrosive chemicals present in the carrier gas. A proper choice of fabric filter typically prevents this problem.

Leaking—This is caused by fine particles penetration or pinholes on the fabric, which is common in fabric filtration. One reason that baghouses are costly to maintain is the difficulty to find and replace even a single leaking bag.

Pressure Drop—Due to the design of FFs, most of the energy consumption to operate the system appears as pressure drop across the bags and ducts. A baghouse key designing factor is to optimize the filter velocity that affects pressure drop (operational cost) and baghouse size (capital cost). Baghouse size is reduced as the filter velocity (or G/C ratio) is increased. However, higher G/C ratios cause higher pressure drops.

Continuous operation—For FFs using shaking and reverser air flow–cleaning methods, the bag filter must be of a multicompartment type, to allow individual units of the bag filter to be successively off-stream during cleaning. In this circumstance, extra cloth area must be considered to ensure that the collection efficiency is not reduced during cleaning periods. However, FFs with pulse jet are not necessary to have extra filter cloth and the baghouse sizes are relatively small, since the pulse jet–cleaning processes are accomplished in a short period of time.

Advantages and disadvantages of FFs

The advantages of FF are: (1) relatively high removal efficiency for fine particles, (2) simple and mature technology in terms of construction and operation, (3) dry collected material disposal, (4) low power consumption. The power consumption of a FF is moderate compared to an ESP.

The disadvantages of FF are: (1) operation limited by flue gas conditions, such as high humidity, temperature, etc., (2) relatively high maintenance costs (the characteristics of the flue gas and the fly ash decide the frequency of FF maintenance requirement, such as bag repairing or replacement, (3) large size equipment, (4) relatively large pressure drop.

As sizes of recent coal-fired power plant are quite large with large volumes of flue gas, FF is not suitable to be applied for PM removal alone. However, due to its capability to capture fine particles, FFs are nowadays integrated with ESPs for ultra-low emissions coal-fired power plants.

3.1.1.3 Key PM control technologies for ultra-low emissions coal-fired power plants

Typically, the overall PM removal efficiency of DESP or FF can be up to 99%+. However, the removal efficiency for fine particles, especially for particles with diameters of 0.1–1 μm, is lower than 90% by using traditional DESPs and FFs [11]. To meet new strict PM regulations, ultra-low emissions coal-fired power plants mainly use three technologies for PM removal:

- Low-Low Temperature ESP (LLT-ESP)
- Wet ESP (WESP)
- Hybrid ESP-Fabric Filter (ESP-FF)

Low-low temperature ESPs (LLT-ESPs)

LLT-ESP is a type of cold-side ESP with lower flue gas inlet temperature (below 100°C, typically 85–95°C) than typical cold-side ESPs (generally 130–180°C). An LLT-ESP requires installing a flue gas heat exchanger before the conventional DESP, to decrease the flue gas temperature entering the ESP, below the acid dew point [12]. Condensed water from the steam turbine then is heated by the hot flue gas. While this process provides for implementation of an energy conservation option, the removal efficiency and the ability to capture SO_3 are significantly improved due to the lower temperature of the flue gas, which decreases the volume of the flue gas entering the ESP, as well as its fly ash resistivity. When the flue gas temperature is reduced by 10°C, the gas volume is decreased by 2.5%–3% [13].

The advantages of LLT-ESP are: (1) low flue gas temperature increases specific collection area due to a more dense flue gas, (2), low flue gas temperature increases water condensate onto the fly ash particles, increasing surface conductivity, (3) low flue gas temperature can extend the life of the ESP due to smaller thermal stresses. All these advantages of flue gas-reduced temperature decrease fly ash resistivity, which raises the PM removal efficiency, reduces the overall size of the unit, and promotes utilization of waste heat from the flue gas at the same time. However, the disadvantages of a lower flue gas temperature is that fouling and corrosion would be caused by sulfuric acid (H_2SO_4), ammonium sulfate $(NH_4)_2SO_4$, and ammonium bisulfate, NH_4HSO_4 (ABS), which are formed in this temperature range.

Wet ESPs (WESPs)

WESP systems have been used to control PM emissions, which are wet, sticky, corrosive, explosive, flammable, or with high resistivity for more than 100 years in industrial settings and for more than 30 years in coal-fired power generation [14]. Typically, a WESP system is used worldwide as individual modular unit in commercial and industrial applications. Due to more restricted regulations for PM emissions from coal-fired power plants, WESPs are popular for installations downstream of wet flue gas desulfurization (WFGD) units, as a final "polishing" stage to remove very fine particulates, sulfuric acid (H_2SO_4), and any other acid mist.

Similar to DESPs, there are also three fundamental steps for WESPs collection processes: particulate charge, particulate collection, and removal of collected particulate. The only difference between DESPs and WESPs is the way to remove collected particulate from the collection electrodes. WESPs require a water flow to be injected or sprayed continually or intermittently to wash collected particles from the collection electrodes into a sump. As a consequence, PM does not accumulate on the collection electrode and no re-entrainment occurs. In this process, a portion of the water may be recycled to reduce total water consumption. The remaining water with collected particles is disposed as sludge. WESPs can handle a wide variety of pollutants and flue gas conditions. As mentioned earlier, a small amount of SO_2 converts to SO_3 in boilers and SCRs, which can further form sulfuric acid passing through the WFGD. In this circumstance, a WESP is well suited for capturing the H_2SO_4 mist. WESPs also have high removal efficiency for fine particles like PM_2, due to the high humidity conditions during collection. WESPs can work at low temperatures, which potentially condenses all water vapor leaving the stack. Water sprays precondition the flue gas and capture some particles. Because the water sprays are located over the top of the electrical fields, the collection plates are continuously wet and continuous washes prevent the build-up of a layer of particulate cake. As a consequence, captured particles are not re-entrained and the concentration of fine PM is reduced. In addition, when burning a low sulfur coal, which produces a high resistivity fly ash, the electrical field does not collapse by using a high power since the voltage applied within a WESP can be higher than in a DESP.

WESPs also have additional advantages over DESPs. No rapping is needed in WESPs with water flow cleaning the collection electrodes. This eliminates emission peaks created by the rapping process in DESPs. In addition, the high humidity in WESP helps to lower the flue gas temperature, which increase the specific collection area (SCA) and lowers fly ash resistivity. WESPs also have moderate energy consumption and a robust performance that is not affected by particles properties, such as resistivity and size distribution. However, water consumption is high and wastewater is generated, making a wastewater treatment system required. Flue gas passing through the WESP is even cooler than for an LLT-ESP, which could be an issue for downstream equipment or stack and may need an additional flue gas heater.

Continuous and intermittent washing of the wet ESP are the most common types of cleaning methods used [6, 15] for WESPs. The major differences between WESPs and DESPs are presented in Table 3.1, which is summarized in Ref. [15]. In summary, the advantages of WESP are: (1) it has no issues with rapping and eliminates re-entrainment created during rapping, (2) there are no back coronas; therefore, higher corona power can be used, (3) the flue gas temperature decreases due to high humidity, which increases fly ash resistivity and specific collection area, which lead to high removal efficiencies, (4) WESPs are compatible and easily integrated with

Table 3.1 Major operational and design differences between WESPs and DESPs [15]

Parameter	WESP	DESP
Flue gas temperature (°C)	45–65	120–450
Flue gas humidity	100% (Saturated)	Typically <10%
Power density	Significantly higher than DESP	Variable with coal sulfur content and ash properties
Particulate resistivity	No impact	Critical design and operation factor
Gas velocity (m/s)	~3	1–1.5
Treatment time (s)	1–5	>10
Re-entrainment	No impact	Important design and operation factor
Construction material	Special metals, plastic with excellent corrosion resistance	Typical mild steel
Water consumption	Substantial and wastewater treatment needed	No water needed

other scrubbing technologies to control other pollutants. The disadvantages of WESP are: (1) WESP increases the complexity of the PM control system and adds water consumption, (2) due to H_2SO_4 formation, WESPs need to be made of corrosion-resistant materials, which are expensive, (3) WESP systems need wastewater or disposal treatment, which adds to the capital and operational expense, (4) higher power consumption, (5) flue gas heat-up may be needed.

Hybrid ESP-fabric filters (ESP-FFs)

New Hybrid ESP-FF systems are supposed to become more viable than the individual technologies by combining the advantages of both ESPs and FFs. The ESP advantages of low-pressure drop and low cost are combined with sorbent injection capabilities and capturing rapping peaks of FFs, which makes Hybrid ESP-FFs one of the effective ways for dust emissions control, especially for $PM_{2.5}$ emission, and required high removal efficiency. Hybrid ESP-FFs operate at the same temperature range of ESPs or FFs. Therefore, Hybrid ESP-FFs can be fabricated by carbon steel, which makes its cost lower than LLT-ESPs and WESPs. In addition, the corrosion issue may also be less serious in these systems. Several types of Hybrid ESP-FF collectors have been developed, including compact hybrid particulate collector (COHPAC), Advanced Hybrid Particulate Collector (AHPC), and TOXECON.

COHPAC can be classified into two types: integrated and separated (Fig. 3.6). In this design, the front ESP collects most of the large-diameter particles and then the rest fine particles are collected in the FFs. The integrated type has the ESP and FF in one built unit, with a direct connection between the two elements. In addition, the integrated type has lower

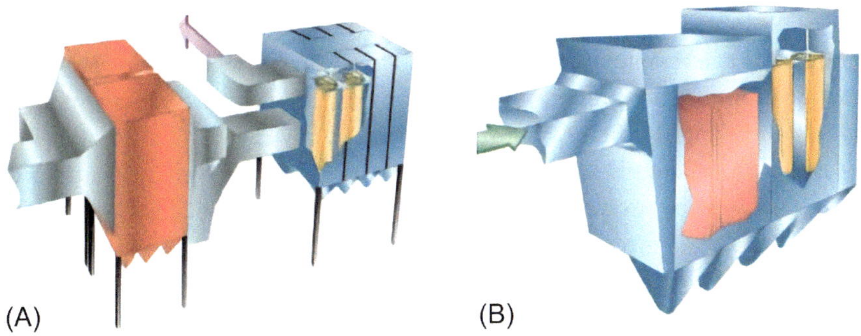

Fig. 3.6 (A) Separated ESP-FF; (B) Integrated ESP-FF [16].

pressure drop but also worse protection to the fabric filter from high-temperature gas and maintenance convenience. In the separated type, the ESP and FF are connected by a duct, and the FF acts as the independent dust collector. Therefore, the integrated type is suitable for new projects, and the separated type is more suited for existing plant retrofits [16].

The other two Hybrid ESP-FF technologies are AHPC and TOXECON. AHPC design consists of an arrangement of filter bags directly between the ESP collection plates, instead of placing the ESP and fabric filter sections in series. The AHPC has a complicated structure in comparison to the COHPAC system. The TOXECON design integrates a sorbent injection system upstream of a FF to the COHPAC. Sorbent is also injected in this design to capture other toxic air pollutants such as mercury, NO_x, SO_3, etc.

3.1.2 Desulfurization technologies

For coal-fired power plants, SO_2 emission can be controlled by a variety of methods that include coal cleaning, in-furnace injection, and post-combustion treatment. The postcombustion SO_2 removal technology is the predominant technology and most used worldwide to control coal-fired power plant SO_2 emissions.

3.1.2.1 Postcombustion FGD technology

Postcombustion FGD technologies are presented in Fig. 3.7. The post-combustion SO_2 removal processes can be classified as once-through systems and regenerable systems, depending on how the sorbent is handled after the desulfurization process. Once-through systems treat the spent sorbent as disposal or utilize it as byproduct. Regenerable systems recycle or regenerate the sorbent back into the system. In the sorbent regenerable process, SO_2 is released and further processed to yield H_2SO_2, raw sulfur, or liquid SO_2. No waste is produced in regenerable systems. Both once-through systems and regenerable systems can be subdivided into wet or dry depending on whether the sorbent state is aqueous or solid. In wet processes, the treated flue gas leaving the FGD system is saturated with H_2O and a liquid slurry waste or byproduct is produced. In dry processes, a dry sorbent is injected. A dry waste or by product is produced with nonsaturated flue gas leaving the system. Currently, regenerable process costs are higher than once-through processes. However, the regenerable process is still an option if space or disposal options are limited and markets for byproduct are not available [17]. For once-through FGD systems, the dry and wet cases can be used in

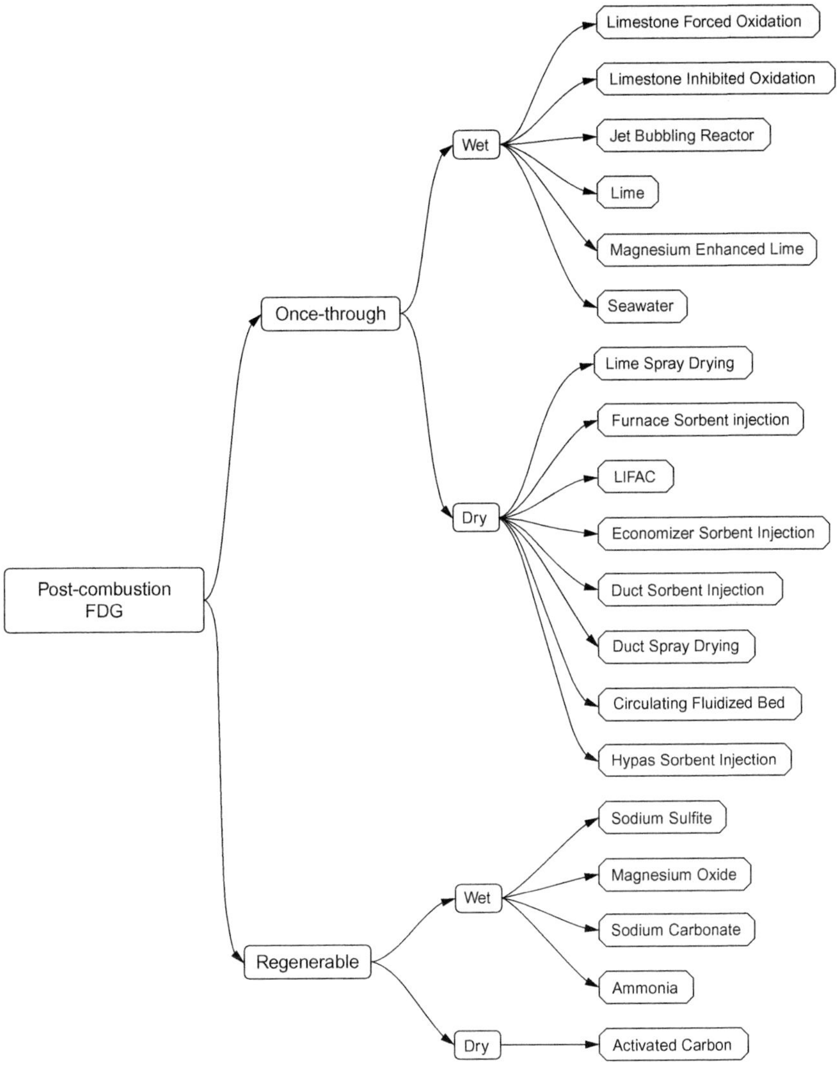

Fig. 3.7 Classification of postcombustion FGD processes [18].

different application arrangements that are cost compatible with desirable removal efficiencies. Limestone-based WFGD is the technology of choice in most applications for small and large coal-fired power plants. The largest dry FGD absorber module in use in the industry is in the range between 250 and 350 MW_e. Considering the relatively less popularity of regenerable FGD processes and dry FGD processes used in coal-fired power plants, this section mainly focuses on once-through WFGD processes. Accordingly, when WFGD/DFGD is mentioned in the remainder of this section, it is in reference to once-through WFGD/DFGD.

3.1.2.2 Key FGD technologies for ultra-low emissions coal-fired power plants

Among desulfurization methods, the once-through WFGD technology is the most-used around the world, with approximately 95% of the installed capacity [19]. WFGDs are often further categorized by reagent types. Limestone is commonly used as the reagent of choice for WFGDs. However, WFGD systems can also be designed for different reagents, including lime, magnesium enhanced lime, ammonia, sodium carbonate, and seawater. In addition, WFGD systems are successfully used for a complete range of coals such as lignite, bituminous, anthracite, and sub-bituminous.

In a conventional WFGD system using limestone as the sorbent, the flue gas enters into a spray tower where limestone slurry is sprayed to absorb SO_2 and produce gypsum for disposal or reuse as a byproduct. The overall reaction is written as:

$$CaCO_3 + SO_2 + 2H_2O + 0.5O_2 \rightarrow CaSO_4 \cdot 2H_2O + CO_2 \quad (3.8)$$

To meet the requirement of single-digit ppm SO_2 emissions level from coal-fired power plants, design modifications to enhance SO_2 reduction have been introduced. Based on the conventional limestone forced oxidation system, which is the most used FGD technology all over the world, different technical routes and modifications have been developed for ultra-low emissions coal-fired power plants, capable of achieving 98% to 99% SO_2 removal efficiencies. This section focuses on the latest state-of-art WFGD designs and configurations for ultra-low emissions coal-fired power plants, including:

- Tray Absorber WFGD Technology
- Dual-loop WFGD Technology
- Double Contact Flow WFGD Technology
- Jet Bubbling Reactor

Tray absorber WFGD technology

A WFGD system combining trays and open spray tower is marketed by Babcock & Wilcox Company (B&W). This tray absorber technology consists of a perforated sheet located directly above the flue gas inlet, which makes the flue gas uniformly distributed through the absorber spray zones. Then, the flue gas and the limestone slurry form a very intense turbulence flow field above

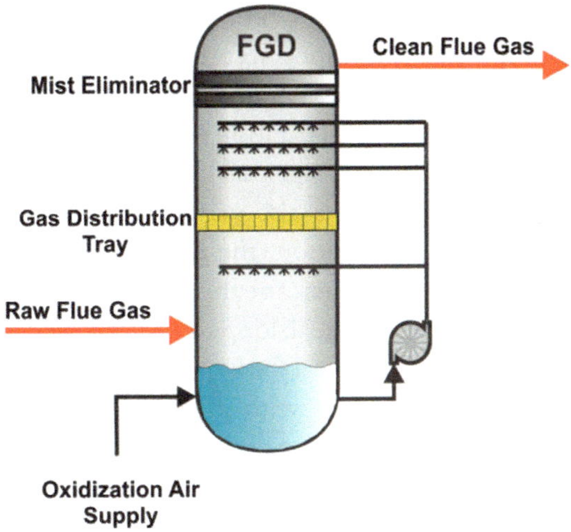

Fig. 3.8 Tray absorber technology developed by Babcock & Wilcox.

the tray, increasing the contacting time between the slurry and the flue gas, which significantly improves SO_2 removal efficiency.

To meet the most restricted SO_2 regulations, B&W has developed a tray tower by adding another level of spray under the tray (Fig. 3.8). In this design, the hot flue gas enters the tower and is cooled down to become saturated with liquid slurry from the spray under the tray and falls from the distribution tray. Then, the saturated flue gas is distributed by the tray and passes through it to react with the slurry from the sprays above the tray. In order to achieve 99% SO_2 removal efficiency, high liquid-to-gas (L/G) ratios are necessary. However, this corresponds to increased pumping capacity, which adds to system power consumption.

Dual-loop WFGD technology

When there is particularly high concentrations of SO_2 in the flue gas, or if especially high SO_2 removal efficiencies are required, the contact zone of the single loop absorber in the limestone-gypsum process could be divided into two loops. The original dual loop WFGD was first developed and patented in the U.S. in 1978 by Babcock Power Environmental Inc. The dual loop WFGD has the capability to maximize SO_2 absorption and reagent utilization at high pHs, while oxidation and gypsum production are maximized at low pHs, with the introduction of the separation of loops. In addition, this design allows to achieve the same level of SO_2 removal efficiency at a lower L/G ratio. The limestone slurry sprays and flue gas are set in a counter-current setup in two stages. The flue gas passes through the first loop, which operates like a single loop absorber. Then, a high concentration of sorbent is used in a second loop. Due to the high excess of absorbent in the second loop, very high levels of separation are achieved, while at the same time the system is well able to manage quick changes in load and changes in flue gas SO_2 concentration. This is also very important for flexible cycling coal-fired unit operation. In addition, the height of the absorber is much lower than the single loop absorber, for the same SO_2 removal efficiency.

As shown in Fig. 3.9, in the dual-loop design, the first loop operates at pH value of 4.0 to 4.5, which is ideal for byproduct gypsum production and oxidation; the second loop works with pH values between 5.8 and 6.4, which is ideal for SO_2 absorption and better reagent utilization. It has been claimed that the first loop can also reduce PM, HCl, and HF in the flue gas, which is beneficial for the second loop to reach higher SO_2 absorption rates. This advantage helps the dual loop WFGD system to reduce the power consumption of the desulfurization process with SO_2 removal efficiency greater than 99%. Due to this more efficient process, operating costs are significantly lower than for single loop absorbers.

Single tower and dual tower, double contact flow WFGD technology

Double contact flow WFGD was originally developed by Mitsubishi Heavy Industries, Ltd. (MHI) [20]. In this design, the flue gas is ducted to the bottom of the absorber where the liquid sorbent slurry is sprayed upward as shown in Fig. 3.10. The flue gas reacts with the slurry as it is sprayed out of special

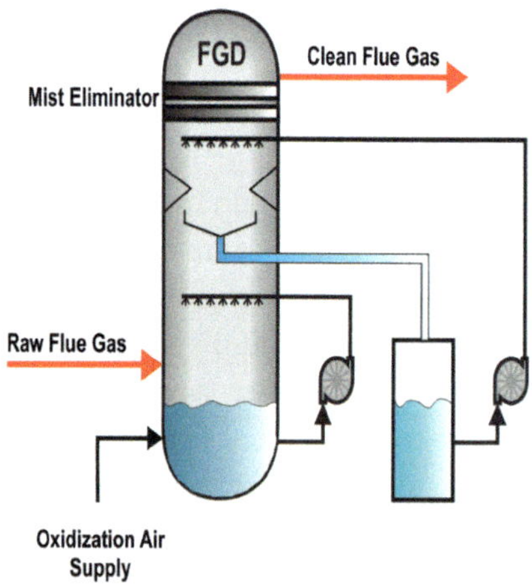

Fig. 3.9 Dual-loop WFGD Tower technology by Babcock & Wilcox.

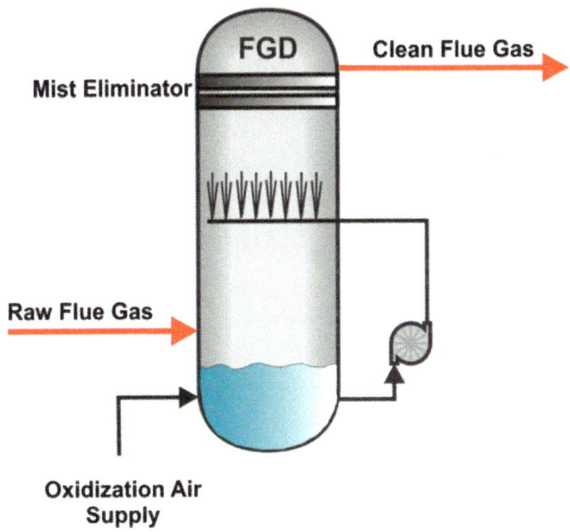

Fig. 3.10 Single tower countercurrent type, double contact flow WFGD by MHI [20].

wear-resistant nozzle on a cocurrent design; then, the liquid slurry falls back down to contact the flue gas again, which flows in a countercurrent fashion. Therefore, the flue gas double contact with the limestone sorbent potentially increases SO_2 removal efficiency. Compared to a conventional scrubber, which is internally filled with packings, porous plate, and multiple spray piping, the contact flow tower is structurally pretty simple, with the spray pipes installed right above the inlet duct. Maintenance for this design is relatively simpler, the cost is lower, and the size is more compact. Since the spray nozzles are just above the inlet duct, there is a reduction in scaling. The advantages of this technology stated by MHI are: (1) SO_2 removal efficiency of more than 99% can be achieved due to the high gas-liquid contact efficiency, (2) easy maintenance: no internal parts other than a single-stage nozzles, (3) negligible scaling.

For larger units, higher sulfur content coal as well as situations requiring higher SO_2 removal efficiencies, a second section of double contact flow scrubbers can be added to achieve the emission reduction target (Fig. 3.11). The inlet tower is referred as a cocurrent tower and the outlet tower is a countercurrent tower. The dual tower configuration can reach SO_2 removal efficiencies of 99%+. However, increased capital and O&M costs are required for the dual tower configuration.

Jet-bubbling reactor WFGD technology

A jet-bubbling reactor (JBR) is based on the same chemistry reaction as the limestone forced oxidation FGD. This process was developed by Chiyoda

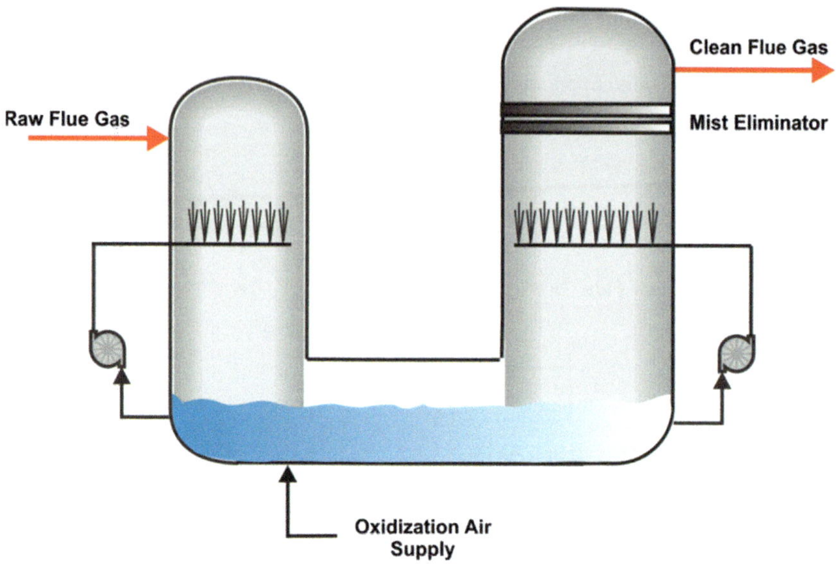

Fig. 3.11 Dual tower double contact flow WFGD by MHI.

Corporation in Japan. The difference of JBR WFGD and other limestone forced oxidation WFGD is the use of a jet-bubbling concept. As shown in Fig. 3.12, the flue gas is cooled by pond water and then is forced down through a series of sparger pipes above the reaction zone, which is immersed in the limestone slurry. Then, the flue gas bubbles upward though the absorber slurry to form a jet-bubbling zone (see Fig. 3.12). This layer is a gas-liquid contact zone to promote SO_2 absorption before going back into the reaction tank. The Chiyoda Thoroughbred 121 (CT-121) WFGD has been demonstrated in the U.S. and Japan and is recognized as a competitive WFGD technology in the power generation market. It has been reported to have SO_2 removal efficiencies higher than 99%. In addition, it can also achieve >90% of particulate removal efficiency and 40%–60% SO_3 removal efficiency [21]. However, these high removal efficiencies are achieved at the expense of increased flue gas pressure drop bubbling through the reactor, which leads to higher ID fan power consumption. This increase of power consumption can be compensated by low recirculating pumping power to feed slurry.

The optimal pH range for conventional spray tower limestone-based WFGD is typically 5.5–6.0. The CT-121 JBR process is designed to operate in a pH range from 3 to 5. The low pH dissolves limestone in the reactor to ensure high SO_2 absorption rates. This also ensures nearly complete reagent

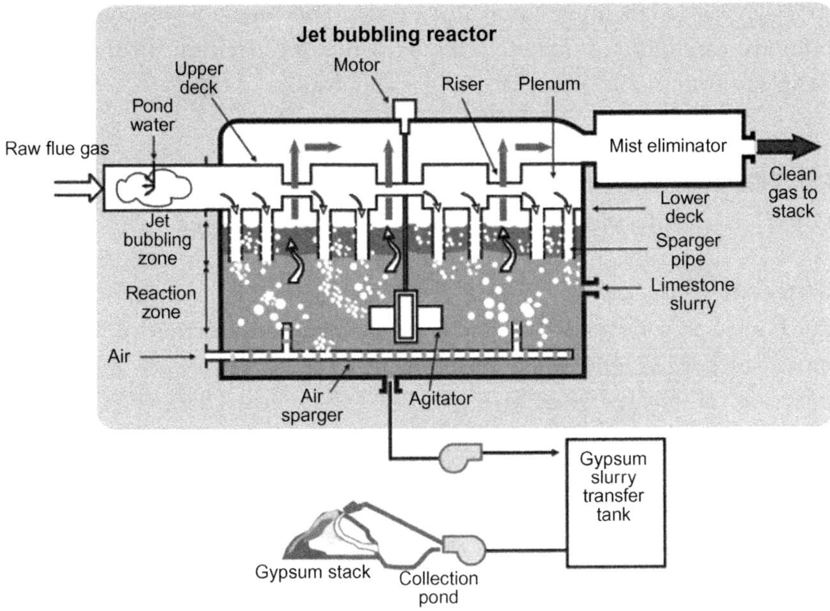

Fig. 3.12 Chiyoga CT-121 JBR WFGD process [22].

utilization. Oxidation of sulfite to sulfate is promoted at this low pH range since the low pH also increases the solubility of natural catalysts such as iron. The JBR WFGD has inherent operating flexibility, which it can be designed for wide operating ranges of flue gas flow rate and SO_2 concentration. This advantage not only well handles cycling power plants, but also meets new changing regulation requirements.

The JBR has evenly spaced sparger pipes, which naturally provides the ability to evenly distribute flue gas to achieve desired SO_2 removal efficiency. The advantages of JBR WFGDs are: (1) high SO_2 removal efficiency, (2) there is no circulation pump needed to feed slurry or very low pumping power, (3) compared to conventional tower WFGD, JBR WFGD has smaller height, (4) able to work at a low pH range and have high reagent utilization, (5) low capital and O&M costs, (6) good flexible operation, (7) good flue gas distribution.

The disadvantages for JBR WFGDs are: (1) high flue gas pressure drop and high ID fan power requirement, (2) relatively wide foot print.

In this section, commercially available WFGD technologies for ultra-low SO_2 emissions in coal-fired power plants are introduced and described. In practice, factors such as flue gas flow distribution, ideal flue gas velocity, spray levels design, mist eliminator design, and selection of liquid-to-gas ratio are carefully considered. This influence the decision on the type of design is more suitable for a particular application.

Dry FGD (DFGD) technology

DFGD is an alternative way to removal SO_2 from flue gas other than WFGD. Typically, DFGD technologies are used to removal SO_2 from sources that burn low- to medium-sulfur coal, and in units with small capacity. Lime Spray Dryer (LSD) and Furnace Sorbent Injection (FSI) are the most widely used DFGD technologies. In some circumstances, dry sorbent injection can be used downstream of FSI; this includes: the LIFAC process, Duct Spray Injection (DSI) and plain sorbent injection in a dedicated Circulating Fluidized Bed (CFB) [18]. Both LSD and CFB require dedicated absorber vessels for SO_2 reactions with sorbents, since larger residence time and contacted surface area increases sorbent utilization. In some cases, additional absorber vessels are not necessary for units with CFB boilers due to the large enough residence time for SO_2 and sorbent reaction. Fig. 3.13 shows feed slurry atomized and injected in a spray dry absorber to absorb SO_2 and then being collected in downstream PM control equipment. As shown in Fig. 3.13, part of the reagent is recycled. The ratio of fresh reagent to recycle reagent is controlled to achieve the desirable SO_2 removal efficiency by maintaining certain level of alkalinity.

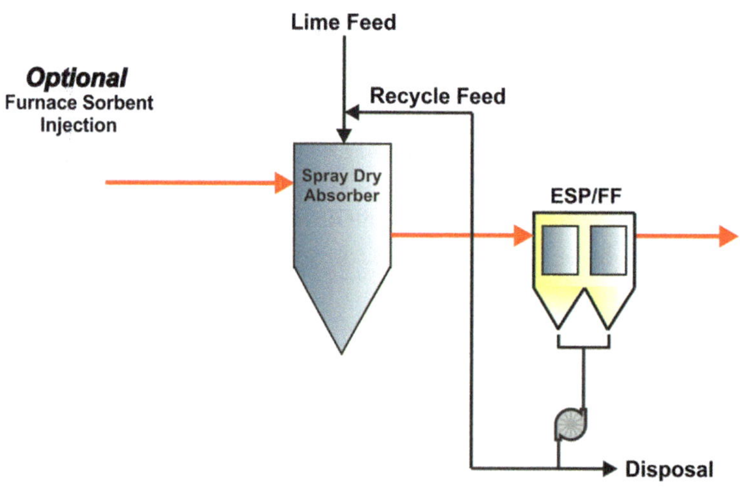

Fig. 3.13 Schematic of DFGD system.

Dry FGD has some advantages and disadvantages, which are included here to compare it to a limestone WFGD [23]. The advantages of a dry FGD are: (1) the absorber vessel can be built of unlined carbon steel, which is cheaper than the material option for the vessel for WFGDs constructed of lined carbon steel or solid alloy. Typically, the capital cost of a dry FGD is lower than a WFGD for units less than 300 MW_e. For units larger than 300 MW_e, modular design for dry FGDs are needed, which are more expensive than WFGD systems, (2) dry waste $CaSO_3$, $CaSO_4$, and $Ca(OH)_2$ are produced, which can be treated by conventional fly ash removal equipment, (3) smaller pumping power and overall low power consumption is required, in comparison to WFGD systems, (4) dry FGD systems require less auxiliary equipment than WFGD systems, which results in lower capital and O&M costs, (5) typically, the flue gas pressure drop passing through the absorber is lower than for WFGD units, 6) SO_3 can also be removed efficiently. WFGD systems can only capture 20%–60% SO_3, (7) the flue gas is not saturated with H_2O after a dry FGD system, which reduces or eliminates wet stacking, (8) there is no liquid waste for dry FGD systems while WFGD systems produce liquid waste.

The disadvantages of a dry FGD are: (1) the largest dry FGD module used in the power industry is 250 to 300 MW_e. For current coal-fired power plant sizes, a series of modules are required, which leads to more ductwork and higher capital cost, (2) a more expensive reagent – lime is used than limestone, which is used in wet processes, (3) reagent utilization is lower than for limestone WFGD systems, to achieve comparable SO_2 removal levels, (4) a large amount of dry waste is produced, which is hard to be reused.

3.1.3 Low-NO_x burners and denitrification

NO_x control in coal-fired power plants significantly depends on coal type, boiler load, and operating conditions. NO_x control is generally achieved by two approaches: combustion control and postcombustion control. For combustion control, this approach is based on the mechanism of NO_x formation discussed at the beginning of this section, there are three avenues for combustion NO_x reduction: (1) decrease flame temperature, (2) reduce residence time in the combustion zone under oxidizing, and (3) create a fuel-rich environment. For the postcombustion control methods, the main approach relies on converting NO_x to N_2 and H_2O through chemical reactions. This process needs introduction of a reagent to react with NO_x either with or without a catalyst.

In coal-fired power plants, the major denitrification technologies are low NO_x burners (LNBs), selective catalytic reduction (SCR), and nonselective catalytic reduction (SNCR). LNB technology is low cost but may not achieve strict emissions limits. The postcombustion NO_x control methods such as SCR and SNCR are introduced in the majority of cases around the world that require stringent NO_x emissions reductions. Both SCR and SNCR technologies let NO_x convert back to N_2 through a series of chemical reactions by using amine-based reagents. In SNCR, the most common reagents are ammonia (NH_3) or urea (NH_2CONH_2). The reagent is injected at the location where the flue gas temperature is between 760°C and 1100°C. The reaction needs sufficient time and proper mixing of flue gas and reagent in this temperature range, which is challenging to control, making it the actual NO_x removal efficiency of SNCR to range between 25% and 75%. Theoretically, SNCR could be able to achieve 90% NO_x reduction. Comparatively, SCR reactions take place at a relative lower temperature range, between 230°C and 450°C. However, the specific temperature range for SCR reactions depends on the type of catalyst formulation. SCR is typically implemented in coal-fired power plants requiring a higher level of NO_x reduction. Theoretically, NO_x removal efficiencies of SCR systems can be designed up close to 100%. However, in practice, commercial coal-fired power plant SCR systems are designed to achieve NO_x removal efficiencies around 90%. SCR is usually combined with other NO_x controls techniques such as LNBs to maintain removal efficiency at high levels [24]. As with SNCR, either NH_3 or urea is used as the NO_x reduction reagent in SCR systems. Urea is generally converted to ammonia before injection.

There are two primary advantages of the SCR process over SNCR: (1) SCR has the higher NO_x reduction efficiency, (2) SCR reactions occur within a relatively lower and wider temperature range. However, SCR has a significantly higher capital and O&M costs than SNCR since large volumes of catalyst used for chemical reactions. In addition, regarding the O&M costs, catalyst replacement represents a large portion of the O&M cost. The most economical and efficient NO_x control solution may be provided by applying multiple denitrification technologies, in a cascade approach. In addition, to achieve emissions goals over a range of fuels and load profile, both combustion optimization and postcombustion control techniques are necessary. Therefore, the most used and effective techniques to control NO_x emission in ultra-low emission coal-fired power plants are LNB and SCR, which are discussed next.

3.1.3.1 Low NO_x burners (LNBs)

Low NO_x combustion can be achieved by either using LNB alone or LNB combined with overfire air (OFA). LNBs, with or without OFA, are the cheapest solution per ton of NO_x removal to reduce 20% to 60% NO_x emissions from uncontrolled levels [25]. Another advantage to use LNBs and OFA is that they are relatively easy to be retrofit for the existing coal-fired utility boiler at relatively low capital cost. In LNBs, air and coal are staged across the burner. There is 75%–85% air supplied to the primary combustion zone with all the required coal. The principle of LNB is creating a fuel-rich environment and lowering the combustion temperature by introducing less air in the primary combustion zone, which restrain NO_x production. The remaining air for complete combustion is introduced in the burner deck as secondary air or even tertiary air after the primary combustion zone. The additional NO_x formation is minimized since the temperature in this zone is not high enough by this staged combustion. The principle of OFA is based on the introduction of remaining combustion air through ports at higher elevations in the boiler, which reduce combustion temperature and reduce additional NO_x production. There are two types of OFA systems: closed-couple overfire air (COFA) and separated overfire air (SOFA). The COFA system basically introduces the remainder secondary air just right above the burners. The SOFA system increases the distance between the primary combustion zone and OFA ports to create a separation zone. In some circumstances, a booster fan is needed to provide sufficient mixing velocity to accomplish higher NO_x reduction rate. For designing efficient OFA systems, parameters like number of corners/elevations, boiler spacing and geometry, and boiler size should be considered.

3.1.3.2 SCR systems

When a low level of NO_x emissions is required by new regulations, SCR systems are the most effective technology to achieve high NO_x removal efficiency. The most common reagent used in SCR system is NH_3. The major difference between SNCR and SCR is that SCR uses a metal-based catalyst with active sites to increase the rate of the reduction reactions. The primary components of the SCR include the NH_3 storage and delivery system, NH_3 injection grid, and the catalyst reactor. In the SCR process, NO_x in the flue gas reacts with injected NH_3 yielding N_2 and H_2O by passing through a catalyst chamber within a specific controlled temperature range (Fig. 3.14). These reactions can be written as:

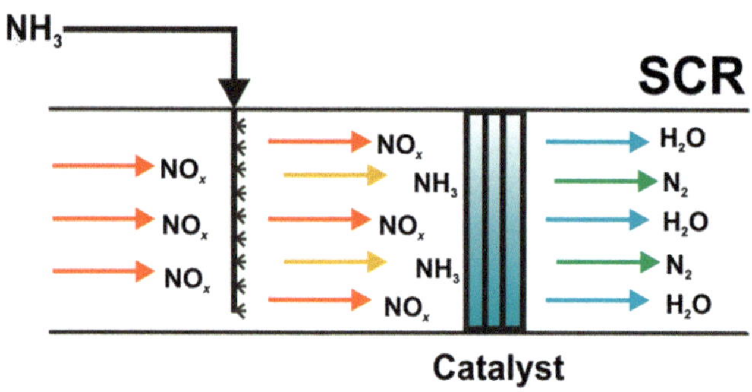

Fig. 3.14 Schematic diagram of typical SCR process.

$$4NO + 4NH_3 + O_2 \rightarrow 4N_2 + 6H_2O \quad (3.9)$$

$$NO + NO_2 + 2NH_3 \rightarrow 2N_2 + 3H_2O \quad (3.10)$$

$$2NO_2 + 4NH_3 + O_2 \rightarrow 3N_2 + 6H_2O \quad (3.11)$$

In the SCR process, the NO_x-NH_3 reactions are highly temperature dependent, which take place in the temperature range between 230°C and 450°C, with optimal operating conditions beyond 360°C, for a mixture of titanium oxide and vanadium oxide. This temperature range determines that the most common location for an SCR system is between the economizer and the air preheater, as presented in Fig. 3.1. This installation is referred to as hot-side, high-dust, since it is located upstream of the ESP and the air preheater (APH). There are also some other different locations downstream of the boiler where SCR systems can be installed. Flue gas temperature and compositions vary with the location of the SCR reactor. It can also be located downstream of the ESP and the APH, which is referred to as cold-side, low-dust. If it is located upstream of the APH and downstream of the ESP, it is called hot-side, low-dust.

- The hot-side, high-dust SCR configuration has an advantage that it does not need a flue gas heat exchanger to heat up the flue gas to the reaction temperature range.
- For the hot-side, low-dust SCR configuration, it is required to be equipped with hot-side ESP, which has been discussed in Section 3.1.1.1.
- The cold-side, low-dust SRC configuration has the longest catalyst life due to working downstream of the ESP and FGD, with low PM and SO_2 levels in the flue gas. However, the disadvantage of this configuration is that the flue gas needs to be heated up, which requires about 2% to 4% of boiler capacity.

Typically, SCR NO_x removal efficiency ranges from 70% to 90% in an SCR system. However, the performance of SCR varies significantly depending on a number of factors, such as coal composition, flue gas compositions, reaction temperature, catalyst formulation and activity, and reagent characteristics, as well as other factors such as prevention of NH_3 slip and SO_3 formation. Operating conditions, which affect SCR efficiency and need to be considered in a good SCR design, are: (1) ammonia slip; (2) catalyst deactivation; (3) flue gas conditions.

1) Ammonia Slip

The right amount of NH_3 injection is one of the key factors to affect SCR performance. Too little injection of NH_3 cannot effectively control NO_x emissions to the desired target level; overinjection of NH_3 will promote undesirable compounds, which can plug and corrode downstream equipment or escape to the environment as hazardous substances. It needs to be mentioned that a small amount of SO_2 in the flue gas is oxidized to SO_3 (Eq. 3.2) in the boiler and over the SCR catalyst. Therefore, these side reactions produce undesirable byproducts, such as ammonium sulfate, $(NH_4)_2SO_4$, and ammonium bisulfate, NH_4HSO_4 (ABS):

$$2NH_3 + SO_3 + H_2O \rightarrow (NH_4)_2SO_4 \tag{3.12}$$

$$NH_3 + SO_3 + H_2O \rightarrow NH_4HSO_4 \tag{3.13}$$

ABS is a sticky and highly acidic substance. The subsequent formation and condensation of ABS, together with the condensation of H_2SO_4 in the air preheater, can cause significant fouling and corrosion issues, which is mainly caused by ammonia slip. In addition, $(NH_4)_2SO_4$ is a fine particle that can lead to increase PM emissions. Therefore, the most important method for reducing NH_3 slip is monitoring and carefully managing NH_3 injection to ensure good NH_3 distribution and mixing with the flue gas. This is also important to maintain catalyst activity at design levels.

2) Catalyst Deactivation

SCR catalysts are typically made of a ceramic material that is a mixture of substrate (titanium oxide) and active components (oxides of vanadium, in some cases, tungsten). The main reasons that cause catalyst deactivation are poisoning, erosion, blinding/plugging, and thermal sintering [26].

Poisoning—Compounds like alkaline metals— Na, K are the prime concern to chemically attach to the catalyst active sites, especially in their water-soluble forms, which can penetrate into the catalyst pores. Gas-phase arsenic (As_2O_3) can condense inside catalyst pores and react with active vanadium sites.

Erosion—Erosion on the catalyst surface can be caused by abrasive solids in the flue gas, such as fly ash, silica (SiO_2), alumina (Al_2O_3).

Blinding/Plugging—Coal with high ash content has the potential of blocking the pores of the catalyst. In addition, Ca in flue gas can react with SO_3 to form $CaSO_4$ to plug the catalyst. In addition, ABS will also deposit on the catalyst surface to blind the active materials.

Thermal sintering—The catalyst activity can permanently be lost due to a change in the pore structure at high flue gas temperatures. Thermal sintering depends on the structure and material of the catalyst. To avoid thermal sintering, flue gas temperature in the SCR has to be controlled carefully.

Aging—The physical and chemical properties of the catalyst pores can change over time to cause the catalyst lose activity.

3) Flue Gas Conditions

The flue gas conditions affect the effectiveness of SCRs significantly, such as flue gas temperature, flue gas velocity, and NO_x concentration. As mentioned earlier, SCR reactions are temperature dependent. At temperatures above the desired range, thermal sintering occurs. At temperatures below the desired range, the reaction rates decrease and ammonia slip happens. Lower flue gas velocity, which also can be expressed as longer residence time, results in higher NO_x removal rate. Flue gas temperature is also related to optimal residence time. When flue gas temperature approaches the desirable temperature range, residence time decreases. At low levels of NO_x, higher concentration of NO_x will result in higher removal rate since the concentration of reactants affects reaction rate. However, if the level of NO_x in the flue gas gets higher than 150 ppm, the performance of SCR does not increase based on a study by the U.S. Environmental Protection Agency (EPA) [27].

In addition, the selection of catalyst is also critical to the operation and performance of an SCR system since the physical and chemical properties of each catalyst should be optimized for different operating conditions. Although postcombustion NO_x reduction technologies are proven and effective, avoiding NO_x formation is always encouraged. The post-combustion NO_x reduction techniques should be always combined with LNBs with/without OFA systems. Furthermore, combustion tuning, balancing air and fuel, as well as controlling fuel finesses are also effective means to achieve cost-effective NO_x reductions. All available NO_x reduction techniques need to be assessed for coal-fired power plants before making any investment decision for NO_x emissions compliance.

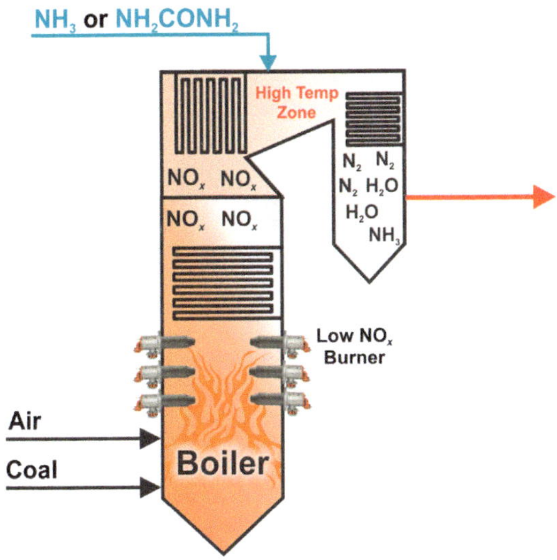

Fig. 3.15 Schematic of SNCR process.

3.1.3.3 SNCR systems

The SNCR technology controls NO_x emissions by injecting a reagent into a specific temperature zone in the boiler. The typical temperature range is from 760°C to 1100°C. This temperature window is critical to the reaction rate of reagent and NO_x. If the flue gas temperature is lower than the expected range, the reaction rate is not large enough to promote full utilization of the reagent for high NO_x emissions reduction. This also results in ammonia slip. On the contrary, excessively higher flue gas temperatures lead to oxidation of the amine in the reagent, creating more NO_x emissions. There is an optimal temperature window at which the reaction rates are optimal for NO_x reduction and the ammonia slip is under control. The reactions in the SNCR and SCR processes are very similar, where a chemical reagent (NH_3 or urea) reacts with NO_x and reduced to N_2 and H_2 (see Fig. 3.15). The SNCR process NO_x removal efficiency can be pushed up to 50%–60%; however, a side problem is high ammonia slip, which is undesirable. The SNCR ratio of reagent to NO_x is typically in the range of 2:1 to 4:1, depending on the reaction temperature and the ammonia slip acceptable limit (typically, <2 ppm).

3.2 Ultra-low emissions technology roadmap for coal-fired power plants

To achieve ultra-low regulatory emissions requirements, techniques used for PM, NO_x, and SO_2 emissions control are necessary to work in a coordinated fashion. In addition, to make a coordinated $PM/SO_2/NO_x$ emission control system work efficiently and effectively, the impact of each type of device on downstream equipment and balance-of-plant should be investigated. In sum, the following factors should be taken into account when planning an emissions control system for a coal-fired power plant.

3.2.1 Coal property

Coal property is very critical during selection of the proper emissions control technology roadmap. It affects flue gas composition, fly ash concentration and properties, and pertinent emissions potential, all of which determines the performance and emissions control equipment and its operation.

3.2.2 SCR impact

SCR technology is successful in removing NO_x in the flue gas. However, the impact of SCR on other emissions control devices needs to be taken into account, such as submicron particle forming, particle cohesion increase as well as SO_3 production. Submicronic particle creation is a recognized issue arising from the use of SCR for NO_x reduction. Submicron particles forming in SCR can refer to $(NH_4)_2SO_4$ and NH_3 due to NH_3 slip. Fine particles are able to increase space charge load as well as change the voltage-current characteristics of an ESP. Particle cohesion is due to formation of ABS, which potentially causes APH fouling and plugging. However, cohesion has a positive effect on ESP efficiency, since it increases the particle size and reduces re-entrainment.

3.2.3 WFGD impact

A WFGD can remove some particulate besides SO_2. However, this process is a function of type of WFGD used, particulate loading at the inlet of the WFGD, and particle size distribution. Therefore, some fine particles and aerosols may be emitted at the outlet of WFGDs, which are possibly carry-over emissions that pass through the scrubber or mist that escape the WFGD mist eliminator. As discussed previously, SO_3 vapor cools down when it passes through the APH and forms H_2SO_4 vapor. When the H_2SO_4 vapor passes through the WFGD, it further cools and forms fine H_2SO_4 aerosol. This aerosol scatters sunlight to create a visible "blue plume." Moreover, the wet flue gas from WFGDs with moisture and gypsum carryover causes "gypsum rain" and condensation in the stack.

The combined impact of different emissions control techniques is summarized in Table 3.2. It is obvious that LNBs, SCR, and WFDG

Table 3.2 Combined effect of different emissions control techniques [13, 28, 29]

Emission	PM	SO_2	SO_3	NO_x
LNB	No interaction	No interaction	No interaction	Reduce 20%–60% NO_x production
SCR	Particles deposit at surface of catalyst to cause fouling and plugging	SO_2 is oxidized to SO_3. Conversion rate is less than 2%		Directly remove NO_x with efficiency >90%
LLT-ESP	Resistivity reduced with lower flue gas temperature. Removal efficiency increases	SO_3 is removed by adhered to particles with efficiency up to 90%		DC corona discharge oxidize NO to NO_2 and use sorbent to absorb NO_2
WESP	Removal particles with high efficiency, especially to fine particles like $PM_{2.5}$	SO_2 reacts with H_2O forming sulfurous acid droplet and being captured	SO_3 reacts with H_2O forming sulfuric acid droplet and being captured	DC corona discharge oxidize NO to NO_2 and use sorbent to absorb NO_2
Hybrid ESP-FF	Suitable with high concentration fly ash, low sulfur content, and high resistivity flue gas with high removal efficiency	No interaction	Partial SO_3 removal	No interaction
WFGD	Some particles are captured, but submicron particles are regenerated	Removal SO_2 with efficiency of 99%+	Removal 30%–60% SO_3	No interaction
DFGD	Sorbent injection introduces particles and ESPs/FFs downstream are required	Removal SO_2 with efficiency of >90%	Removal SO_3 with efficiency up to 98%	No interaction

technologies will be predominant and indispensable for NO_x and SO_2 reduction, although different modifications and improvements may be required. In addition, using SCRs and WFGDs may not effectively provide enough removal of other toxic pollutants than NO_x and SO_2. The effective capability of fine particle and aerosol removal of PM capturing devices makes them very essential in compliance plans for ultra-low emission coal-fired power plants. Finally, the cost of ultra-low emissions control technologies for coal-fired power plants is an important consideration factor in their selection. Cost varies greatly with different geographical areas, construction time, and technology availability and maturity.

3.2.4 Emissions control technology based on DESP and WESP

In this configuration (Fig. 3.16), the DESP is the first and main control device to remove the heaviest load and most of the particulates, while the WESP is the last stage control device before the stack and act as a final "polishing" device. Both DESPs and WESPs are able to remove PM with high efficiency. The combination of DESP and WESP can maximize their effectiveness with goal of achieving ultra-low emissions.

DESPs work with high PM loading at a higher temperature range (120–450°C, in this particular case, which is using a cold-side DESP at 120–180°C). In addition, DESPs fabricated from mild carbon steel have low-to-moderate capital cost compared to WESPs. However, DESPs are not capable of removing fine particles (such as $PM_{2.5}$) with efficiencies up to 99%. SO_3/H_2SO_4 cannot be removed by DESPs unless sorbents are injected. Furthermore, DESPs also have resistivity, back corona, and re-entrainment issues, which potentially decrease removal efficiency. Under

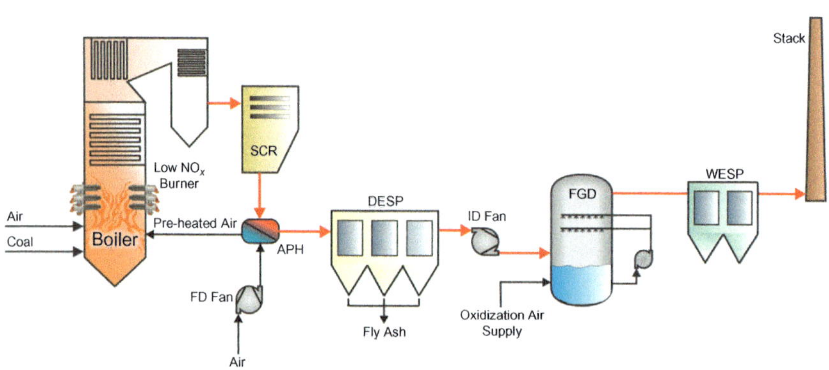

Fig. 3.16 DESP- and WESP-based emissions control.

this scenario, WESPs are installed downstream of WFGDs. The flue gas enters the WESP at a saturated state. The WESPs are placed at the tail end to collect $PM_{2.5}$, H_2SO_4 mist, and other liquid droplets remaining in the flue gas after the WFGDs. Typically, WFGDs can remove 20%–60% SO_3 depending on the types of WFGD. SO_3 combines with moisture to form H_2SO_4 aerosols with a diameter of 1–20 μm, which are difficult to capture by spray droplets in WFGDs. These aerosols then move to the WESP to be ionized, and migrate to the collection electrodes to be captured. NH_3, HCl, and mercury are similar to H_2SO_4 and can be collected in WESPs.

3.2.5 Emissions control technology based on LTT-ESP

As previously mentioned, a heat exchanger is installed upstream of LLT-ESPs to cool the flue gas down to 85–95°C, and at same time this heat exchanger transfers heat to the condensed water from the steam turbines, which can increase unit heat rate (Fig. 3.17). Due to the fact that the temperature is below the H_2SO_4 dew point, most of SO_3 forms H_2SO_4 mist in the LLT-ESP package and adheres onto the particulate, which significantly decreases the resistivity in the LLT-ESP. The lower resistivity can also prevent back corona from occurring. The decrease in fly ash resistivity potentially results in re-entrainment. The lower flue gas temperature also makes the flue gas denser decreasing flue gas velocity. Residence time and SCA in the LTT-ESP arrangement increase, which raises collection efficiency. PM concentration in stack gas of 10 mg/m^3 and SO_3 removal rate of 80% can be achieved by this technology arrangement.

Additionally, a second heat exchanger may be installed prior to the stack to heat up the flue gas, which eliminates the impact of condensed acid on the

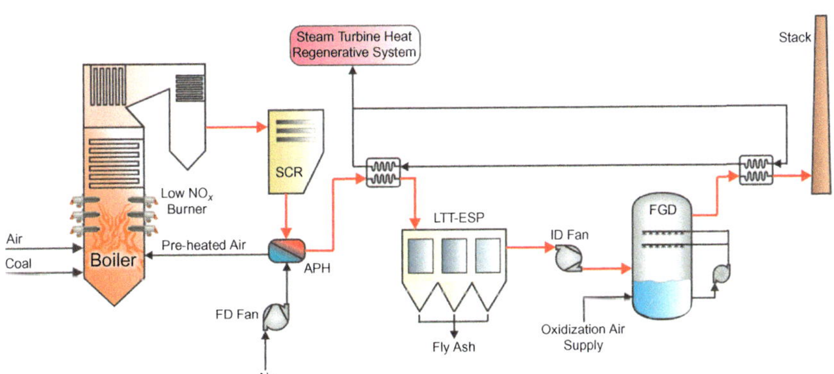

Fig. 3.17 LTT-ESP-based emissions control.

stack. However, corrosion issues both in heat exchangers and LLT-ESP should be carefully examined.

3.2.6 Emissions control technology based on hybrid ESP-FF

This technology arrangement is based on a Hybrid ESP-FF PM collector, which combines an ESP and a FF (Fig. 3.18). In this system, the concentration at the FF inlet is significantly reduced since the ESP is upstream of it, removing large diameter particles. In addition, compared to using a FF alone, pressure drop is also reduced. With both advantages of ESPs and FFs, Hybrid ESP-FFs are capable of removing fine particulate with $PM_{2.5}$ removal efficiency of up to 98%–99% [13]. Since most of the large particles are removed by the ESP, the FF potentially has longer working life than if it was standing along, due to lower abrasion. Another advantage of the hybrid system is operational flexibility. This system has less sensitivity to coal properties and fly ash composition, which can ensure continued stack gas emissions, despite fluctuations in coal and fly ash compositions.

Similar to other configurations, the location of the hybrid collector is in between the APH and WFGD. The SO_3/H_2SO_4 can only be partially removal by the Hybrid ESP-FF system. To meet the requirement of ultra-low emissions, an option could be to install the WESP downstream of the WFGD to achieve desirable low emissions limits.

3.2.7 Emissions control technology based on dry FGD (DFGS)

A schematic of Lime Spray Dryer (LSD) with optional Furnace Sorbent Injection (FSI) is shown in Fig. 3.19. The lime slurry is finely distributed by rotary atomizers or nozzles into the flue gas. As the slurry mixed with

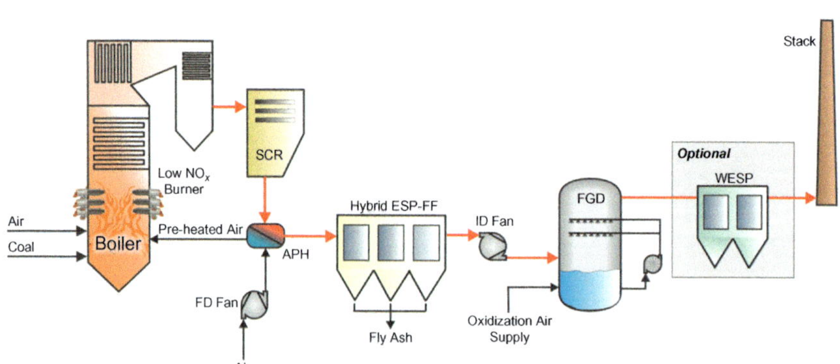

Fig. 3.18 Hybrid ESP-FF-based emissions control.

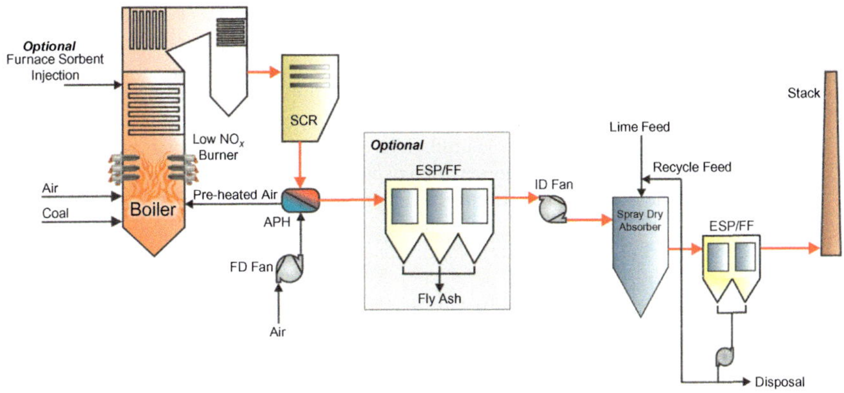

Fig. 3.19 DFGD-based emissions control.

hot flue gas, some water is evaporated. The remainder water in the sorbent promotes the reaction with SO_2. The desulfurized flue gas, along with fly ash, reaction products, and unreacted lime are collected by an ESP/FF downstream for further disposal, and they are used as byproduct or recycled feeding into the dry scrubber. As mentioned previously, the LSD system is not suitable to handle units with capacity larger than 350 MW_e. Multiple absorbers may be required for large combustion units, which may significantly increase capital and O&M costs. Due to the lower L/G ratio than WFGD system use, the SO_2 removal efficiency of LSF systems is lower than the efficiency of WFGD systems. In addition, no wet waste or waste water treatments are required for DFGD systems.

The FSI system requires the flue gas temperature be higher enough to decompose the sorbent into porous solids with high surface area to enhance the absorption rate. The typical sorbents used in FSI are calcium- and sodium-based alkaline reagents. The standalone sorbent injection system can achieve 50%–60% for calcium-based sorbents and up to 80% for sodium-based sorbents [18]. The FSI system is suitable to be integrated with the LSD to achieve high SO_2 removal efficiency. If stricter regulation is promulgated, WESP would be an option for stack to control of fine particles and SO_3/H_2SO_4 emissions. However, the capital and O&M cost will increase as well.

References

[1] Energy Information Administration (EIA). International Energy Outlook 2017, 2017.
[2] D.J. Casagrande, Sulphur in Peat and Coal, vol. 32(1), Geological Society, London, 1987, pp. 87–105. Special Publications.

[3] US EPA (2003). Air Pollution Technology Fact Sheet, Dry Electrostatic Precipitator (ESP)—Wire-Plate Type. EPA-452/F-03-028. United States Environmental Protection Agency.

[4] H. Brauer, Y.B.G. Varma, Air Pollution Control Equipment, Springer Science & Business Media, 2012.

[5] V. Arrondel, G. Bacchiega, M. Hamlil, The electrostatic precipitator external parameters at the heart of dust collection efficiency performance: coal characteristics, combustion quality and SCR chemical process, Int. J. Plasma Environ. Sci. Technol. 6 (2012) 25–32.

[6] K. Nicol, Recent Developments in Particulate Control, vol. 1, IEA Clean Coal Centre, 2013, pp. 29–30.

[7] D. Drehmel, Developments in Particulate Control for Coal-fired Power Plants, U.S. Environmental Protection Agency, Washington, D.C., 2002 EPA/600/J-80/092 (NTIS PB81122707).

[8] G. Bruce, Clean Coal Engineering Technology, Elsevier Science & Technology, 2016.

[9] R. Zevenhoven, P. Kilpinen, Control of Pollutants in Flue Gases and Fuel Gases, Helsinki University of Technology, Espoo, Finland, 2001.

[10] US EPA, Air Pollution Technology Fact Sheet, Fabric Filter—Pulse-Jet Cleaned Type, EPA-452/F-03-025, United States Environmental Protection Agency, 2003.

[11] J. Zhuo, Formation Mechanism of Submicron Particulate Matters During Pulverized Coal Combustion [Dissertation], Tsinghua University, Beijing, 1998.

[12] A. Bäck, Enhancing ESP efficiency for high resistivity fly ash by reducing the flue gas temperature, in: Electrostatic Precipitation, Springer, Berlin, Heidelberg, 2009, pp. 406–411.

[13] W.Z. Shi, M.M. Yang, X.H. Zhang, S.Q. Li, Q. Yao, Ultra-low emission technical route of coal-fired power plants and the cooperative removal, Proc. CSEE 36 (16) (2016) 4308–4318.

[14] A.L. Moretti, C.S. Jones, Power-Gen Asia, Advanced Emissions Control Technologies for Coal-Fired Power Plants, Babcock & Wilox Power Generation Group, Bangkok, Thailand, 2012. Technical Paper, BR-1886.

[15] A.A. Silva, A.L. Moretti, T. Liu, et al., Wet electrostatic Precipitator (ESP) Designs to Achieve Ultra-Low Particulate Emissions, Power-Gen Asia, Bangkok, Thailand, 2015.

[16] Y.P. Yao, X.Y. Zhao, Application of Electrostatic Fabric Hybrid Particulate Collector, Electrostatic Precipitation, Springer, Berlin, Heidelberg, 2009, pp. 482–484.

[17] US EPA, Air Pollution Technology Fact Sheet, Flue Gas Desulfurization (FGD)—Wet, Spray Dry, and Dry Scrubbers, EPA-452/F-03-034, *United States Environmental Protection Agency*, 2003.

[18] R.K. Srivastava, W. Jozewicz, Flue gas desulfurization: the state of the art, J. Air Waste Manage. Assoc. 51 (12) (2001) 1676–1688.

[19] D.H. Zhang, Y. Zhuang, R.R. Zhu, K.W. Liu, Ultra-low air pollutant control technologies for coal-fired flue gas and its economic analysis, Electric Power Construction 36 (5) (2015) 125–130.

[20] Y. Nakayama, N. Kojima, T. Oishi, S. Okino, M. Oniduka, Advanced technology of flue gas desulfurization (FGD), Mitsubishi Heavy Ind. Tech. Rev. 41 (4) (2004) 1–4.

[21] M. Rübner-Petersen, F. Pollastro, Desulphurization technology: Jet Bubbling Reactor CT-121, in: ATI $DeNO_x$–$DeSO_x$ Conference, Milan, 2005, pp. 20–21.

[22] U.S. Department of Energy, Demonstration of Innovative Applications of Technology for the CT-121 FGD Process—A DOE Assessment, DOE/NETL-2002/1177, Morgantown, 2002.

[23] Sargent & Lundy LLC. Dry Flue Gas Desulphurization Technology Evaluation. Project Number 11311-000. Prepared for National Lime Association, 2002.

[24] Sorrels, J. L., D. D. Randall, K.S. Schaffner, and C.R. Fry. Selective Catalytic Reduction, 2015.

[25] Oxides N. Why and How They Are Controlled. Clean Air Technology Center (MD-12), US EPA Technical Bulletin No. EPA-456/F-99-006R. http://www.epa.gov/ttncatc1/dir1/fnoxdoc.pdf 1999.
[26] C.H. Bartholomew, Mechanisms of catalyst deactivation, Appl. Catal. A Gen. 212 (1-2) (2001) 17–60.
[27] US EPA, Air Pollution Technology Fact Sheet, Selective Catalytic Reduction (SCR), EPA-452/F-03-032, *United States Environmental Protection Agency*, 2003.
[28] C.-L. Yang, Pulse-energized wet tubular electrostatic precipitator for NO_x emission control, Environ. Prog. Sustain. Energy 18 (2) (1999) 80–86.
[29] Retrieve from: https://www.aldenlab.com/Portals/0/Documents/Webinars/scrubber-all-slides.pdf.

CHAPTER FOUR

Cases of ultra-low emission coal-fired power plants

Tao Wang, Yongsheng Zhang
School of Energy, Power and Mechanical Engineering, North China Electric Power University, Beijing, China

In Chapter 3, four kinds of ultra-low emissions technology roadmaps for coal-fired power plants are proposed. They are Emissions Control Technology Based on DESP and WESP, LTT-ESP, Hybrid ESP-FF, and Dry FGD (DFGS).

This chapter offers a detailed description and illustration of the technical details and pollutant emission scenarios in connection with different ultra-low emission technology roadmaps as currently adopted by coal-fired power plants in China. This chapter covers only the discussion on pollutants, including PM, SO_2, and NO_x, while the discussions on other pollutants will be made in the following chapters.

The data in this chapter were gathered from online instruments belonging to different power plants, as shown in Table 4.1 with details.

4.1 Emissions control technology based on DESP and WESP

4.1.1 Enhanced PM removal by DESP

4.1.1.1 Wangqu Power Plant

(1) Introduction to the power units before upgrade

In this power plant, two supercritical coal-fired units of 2×600 MW were installed, called Units 1 and Unit 2. They belonged to the large-scale pit and fire thermal power project that was jointly accomplished by Shandong and Shanxi province.

PM removal system

The original electrostatic precipitators as installed on Units 1 and 2 belonged to a double column and double chamber five electric fields ESP, which was put into operation in August 2006.

Advances in Ultra-low Emission Control Technologies for Coal-Fired Power Plants
https://doi.org/10.1016/B978-0-08-102418-8.00004-8

Table 4.1 Monitoring methods and instruments information of ultra-low emission power plants

	PM				SO_2				NO_x			
Power plant	**Detection principle**	**Range (mg/m^3)**	**Model**	**Manufacturer**	**Detection principle**	**Range (mg/m^3)**	**Model**	**Factory**	**Detection principle**	**Range (mg/m^3)**	**Model**	**Manufacturer**
Jiaozuo	Laser-based forward scatter	0–20	181WS	PCME, UK	NDIR	0–75	ABB 2040	ABB	NDIR	0–200	ABB 2040	ABB
Qinhuangdao	Scattering method	0–100	MODEL2030	Beijing SDL Technology	NDIR technique	0–150	ULTRAMAT23	Siemens, Germany	NDIR	0–200	ULTRAMAT23	Siemens, Germany
Sanhe	Filter weighing method	0–20	3012H-D	Qingdao Lonying Environmental Technology	Infrared method	–	DX4000	Finland GASMET	Infrared method	–	DX4000	GASMET, Finland
Huayuan	Laser-induced breakdown spectroscopy	0–200	FW320	Sick Maihak	Infrared method	0–250	S710	Sick Maihak	Infrared method	0–250	S710	Sick Maihak
Yuanyanghu	Laser-based backscatter	0–100	MODEL2030	Beijing SDL Technology	Infrared method	0–500	ULTRAMAT 23	Siemens, Germany	Infrared method	0–400	ULTRAMAT 23	Siemens, Germany
Hequ	Laser-based forward scatter	0–100	MD6000	Nanjing Bory-Durag	NDIR	0–1300	ULTRAMAT23 Gas Analyzer	Siemens, Germany	NDIR	0–1250	ULTRAMAT23 Gas Analyzer	Siemens, Germany
Dagang	Laser-based backscatter	0–50	RBV-DUST	Shenzhen Rainbow Valley	NDIR	0–5000	SMC-9021	Sike McHack	NDIR	0–1000	U23	Beijing SDL Technology Co., Ltd.
Wangqu	Laser emission	0–10/100	MD6000	Nanjing Bory	Ultraviolet fluorescence	0–100/500	MODEL200 43i	Thermo Fisher Scientific	Chemiluminescence	0–150	MODEL200 42i	Thermo Fisher Scientific
Wanzhou	ProScatter™—Forward Scatter	0–30	D-R 820F	Durag	Ultraviolet method	0–100	PA200	Chongqing ChuanYi Automation	Ultraviolet method	0–100	PA200	Chongqing ChuanYi Automation
Hequ CFB	ProScatter™—Forward Scatter	0–30	SBF800	SHBAIF	NDIR	0–100	GCEM4100	Mandrake	NDIR	0–154	GCEM4100	Mandrake
Midong CFB	ProScatter™ – Forward Scatter	0∼15 ∼50; double range	MD6000 particle continuous monitoring system	Nanjing Bory	NDIR	0∼100 ∼300; double range	AO2020 Gas Analyzer	ABB	NDIR	0∼100 ∼300; double range	AO2020 Gas Analyzer	ABB

Desulfurization system

Onto these 2 × 600 MW coal-fired power units, 2 sets of wet limestone gypsum flue gas desulfurization equipment and one set of desulfurization general system were installed. Unit 1's desulfurization system was officially put into operation on September 1, 2006, and Unit 2's desulfurization system on September 15, 2006.

Denitrification system

Units 1 and 2 adopted the selective catalytic reduction denitrification (SCR) technique, with the design denitrification efficiency rate being no less than 80%, the ammonia escape rate was no more than 3ppm, and the SO_2/SO3 conversion rate less than 1%, while the SCR inlet flue gas temperature remained at 300–400°C. Both units were put into operation in April 2010, with the original design of a 2-layer plate catalyst and reserved spare layer.

(2) Roadmaps of the upgrade

Due to the reason that the originally designed emission concentration condition could not meet the ultra-low emission requirements, the power plants implemented ultra-low emission upgrades to these two units in December 2013 and July 2014 separately.

With respect to the electrostatic precipitator, its original single-phase power supply system was upgraded to a three-phase high efficiency power supply, its low-voltage control system was also upgraded, and its main unit was inspected and refurbished. After these upgrades, the performance requirements to the electrostatic precipitators of these two units were at the PM concentration and the stack outlet was less than 20 mg/m^3 when the PM concentration at inlet was no more than 27.4 g/m^3, and the electrostatic precipitator removal efficiency rate was not less than 99.9% when the units were running at the rated load.

As to desulfurization, high-efficiency roof ridge demisters were equipped to work collaboratively for the PM removal purpose. The original flat demister was replaced with the high-efficiency roof ridge defogger module with a three-tier layout arrangement. By this way, performance of the demister was improved and the PM concentration at the stack outlet of the absorption system was reduced. The PM emission concentration at the stack inlet was consistently controlled within 3 mg/m^3 during operation running.

Furthermore, the SO_2 removal system upgrade adopted a single-absorber single-loop limestone gypsum wet desulfurization technology roadmap, featuring an external slurry tank and auxiliary spray layer technology.

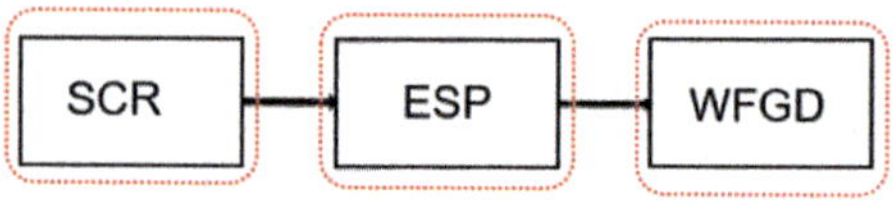

Fig. 4.1 Upgrade technology roadmaps of Wangqu Power Plant.

Both Unit 1 and Unit 2 catalysts were replaced with two-layer honeycomb catalysts. And the discharged plate catalyst was assembled into one layer, changed into the 2+1 mode. In the meantime, upgrades were implemented to the low-NO_x burners. The technology roadmaps concerning the power units before and after the upgrade in Wangqu Power Plant are as follows (Fig. 4.1).

(3) Coal quality

Currently, a burning method of blending lean coal with bituminous coal prevailed in Wangqu Power Plant, with a blending ratio of about 1:1. In the future, lean coal blending ratio would be further reduced continuously. The coal quality was tested from January 2015 to December 2016 separately. And the condition of the specific coal quality during this period was tabulated, based on the year as shown in Table 4.2, and based on the month as shown in the following diagrams too (Fig. 4.2).

It is indicated in the above figure that the average value of total moisture content on the as-received basis was 10.55%, with limited moving range; the average value of the base ash on as-received basis was 19.69%, with a certain amount of fluctuation; the average value of the base sulfur on as-received basis was 0.34%, which indicated almost stability; the average value of the base volatiles on as-received basis was 27.11%, which, though indicating an obvious fluctuating change, did not show any significant changes on the overall data.

Table 4.2 Coal quality condition analysis in Wang Qu power plant

Time	Coal type	Moisture (%)	Ash (%)	Sulfur (%)	Volatile matter (%)	Low calorific value (MJ/Kg)
2015	Blended coal (bituminous coal+lean coal)	10.51	18.96	0.32	27.08	22.11
2016	Blended coal (bituminous coal+lean coal)	10.60	20.49	0.35	27.14	22.37
Mean	Blended coal (bituminous coal+lean coal)	10.55	19.69	0.34	27.11	22.23

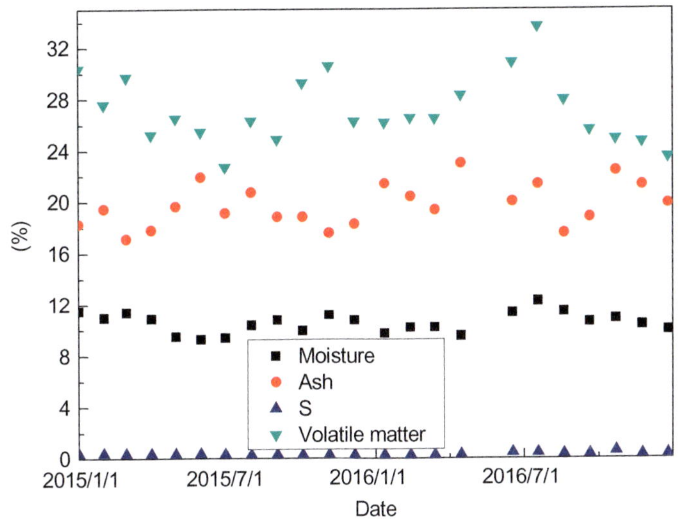

Fig. 4.2 Coal quality analysis in Wang Qu Power Plant.

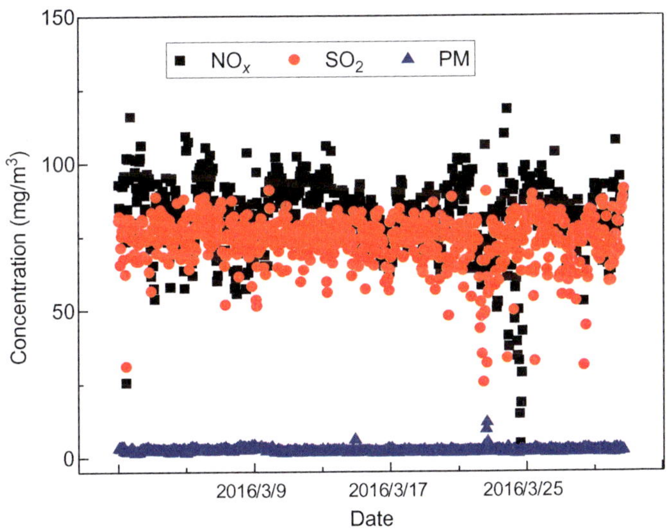

Fig. 4.3 Unit 1 emission data before ultra-low emission upgrade.

(4) Emission status before and after ultra-low emission implementation
Figs. 4.3 and 4.4 show the emission characteristics of PM, SO_2, and NO_x emissions in flue gas at different times before and after ultra-low emission upgrade to Unit 1 in Wangqu Power Plant. The emission concentration

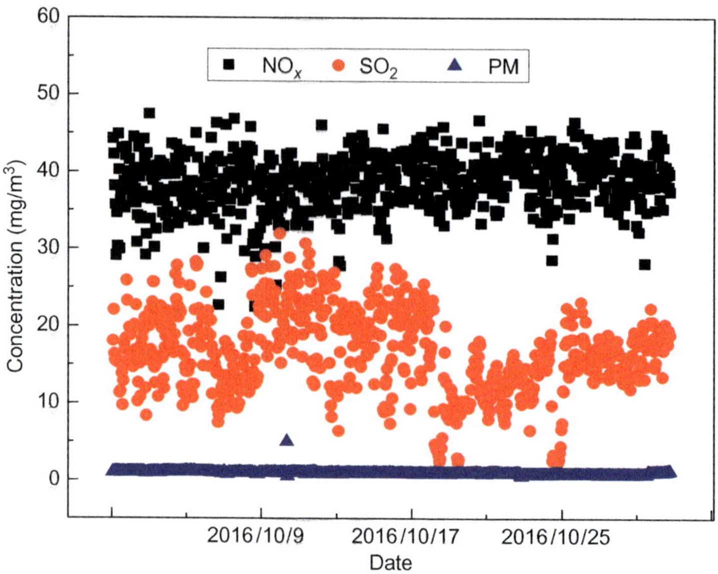

Fig. 4.4 Unit 1 emission data after ultra-low emission upgrade.

of NO_x was mostly at 50–100 mg/m^3 before the upgrade of the power units, among which there were three sets of data far exceeding 200.00 mg/m^3, indicating a rather serious scenario. SO_2 concentration was mostly running between 50 and 75 mg/m^3. The PM concentration was within the range of 0∼5 mg/m^3, with some data still exceeding 10 mg/m^3 on some individual occasions. After the ultra-low emission upgrade, Unit 1's NO_x emission concentration showed a significant cutback, mostly running between 35 and 45 mg/m^3. Even its peak value remained only at 47.45 mg/m^3, lower than the ultra-low emission standard of 50 mg/m^3. The emission concentration of SO_2 also became evidently lower, mostly within the range of 10 to 25 mg/m^3, with the maximum value being only at 31.96 mg/m^3, less than the ultra-low emission standard of 35 mg/m^3. The PM concentration was normally less than 5 mg/m^3, with the peak value being at 5.02 mg/m^3, which could consistently meet the ultra-low emission requirements.

4.1.1.2 Wanzhou Power Plant

(1) Introduction to the power units and their ultra-low emissions implementation

The two 2 × 1050 MW ultra-supercritical coal-fired power units were newly built in Wanzhou Power Plant. They were simultaneously furnished with advanced environmental protection equipment to comply with the ultra-low emission requirements, including the desulfurization system, denitrification system, and electrostatic precipitator PM removal system. These two units were put into operation for power generation on February 9, 2015 and September 18, 2015.

PM removal system

Two sets of 3035 t/h boilers were equipped with two electrostatic precipitators (one for each set). They were high-voltage electrostatic precipitators fitted with high-pressure ESP with three chambers and five electric fields, installed with the high-voltage static electricity electrostatic precipitators + the high-frequency power supply + the three-phase power supply + the tube bundle PM collection device, so as to achieve the ultra-low PM emission goals. The concentration at the stack outlet of the electrostatic precipitators was designed as 15 mg/m^3. Furthermore, on the basis of the foregoing content, a defogging and PM removal device of the centrifugal cyclone tube was additionally installed in the desulfurization absorption system, through which the removal of fine PM particles in coal-fired flue gas was to be achieved. The working principle is explained in this chapter. In the upper part of the desulfurization absorption system, there were a large number of fine droplets contained in the low-temperature saturated pure flue gas. When the fine droplets were in the high-speed motion condition, the probability of PM particles to collide with the droplets increased, resulting in easier condensation of droplets with PM particles and, moreover, achieving the capture of extremely small-sized PM particles. By this way, it was expected that the PM ultra-low emission targets could therefore be achieved.

Desulfurization system

As to the two sets of wet limestone gypsum flue gas desulfurization equipment, the high-efficiency spin coupled desulfurization absorption system and demisters were installed. The overall operating condition of the current desulfurization equipment was quite satisfactory, with an annual operating rate of 100%, showing 0.45% of sulfur content in coal, a desulfurization efficiency of over 97.1%, and a SO_2 emission concentration of $\leq$35 mg/m^3.

Denitrification system

Inside the boiler, a low-NO_x swirling burner was adopted. And the selective catalytic reduction (SCR) denitrification process was adopted in the flue at the same time. Also, the urea pyrolysis process was employed in connection with the reducing agent in this process. Its denitrification efficiency was designed at 85%, with $NO_x \leq 50$ mg/Nm^3 at the stack outlet. The flow chart shows the condition of Wanzhou Power Plant before and after the upgrade as follows (Fig. 4.5).

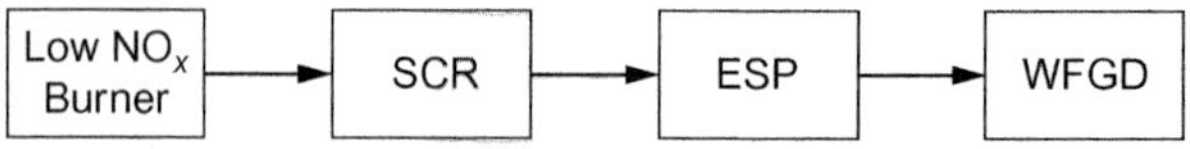

Fig. 4.5 Roadmaps of Wanzhou Power Plant.

Table 4.3 Coal quality analysis in Wanzhou Power Plant

Time	Coal type	Moisture (%)	Ash (%)	Sulfur (%)	Volatile (%)	Low calorific value (MJ/kg)
November 2015	Bituminous	18.02	9.68	0.41	26.11	20.63
November 2016	Bituminous	14.7	11.84	0.38	26.75	21.68
December 2016	Bituminous	14.3	11.11	0.36	26.67	21.44
January 2017	Bituminous	16	10.12	0.39	26.60	21.10

(2) Coal quality

The coal used by the two coal-fired power units originated from Shenhua Ningdong Coal Plant. Tests were carried out for coal quality in November 2015, November and December, 2016, and January 2017. The specific coal quality is shown in Table 4.3.

From the table, it can be seen that Shenhua Ningdong coal's dry ash-free base volatiles as used in Wanzhou Power Plant was about 34%, thus making it the medium-high volatile bituminous coal; the total moisture was around 15%, thus ranked medium-high full-water coal in the whole moisture classification; the calorific value on as-received base was above 20 MJ/kg, thus the low calorific value coal among the bituminous coal and anthracite; the base ash on as-received base was about 10%, thus the low ash coal in the ash classification of thermal coal; the sulfur content was about 0.5%, thus the low-sulfur coal in the sulfur classification of anthracite and bituminous coal; and finally, the grindability coefficient was around 60, thus accordingly making it as medium-grindable coal.

It can be seen from the above classification that Ningdong coal used in Chongqing Wanzhou Power Plant had a higher volatile matter and a higher calorific value among the coals used in coal-fired power plants. This kind of coal can burn easily, by having a low level of ash and a low level of sulfur content. Therefore, it tends to impose less pressure on desulfurization equipment. In general, Ningdong coal used in Wanzhou Power Plant had more advantages, because of its supreme quality, thus imposing much less pressure on the work of the flue gas pollutants removal.

(3) Emission after implementation of ultra-low emission

Both Figs. 4.6 and 4.7 illustrate the emissions characteristics of PM, SO_2, and NO_x of Units 1 and 2 at different times succeeding the

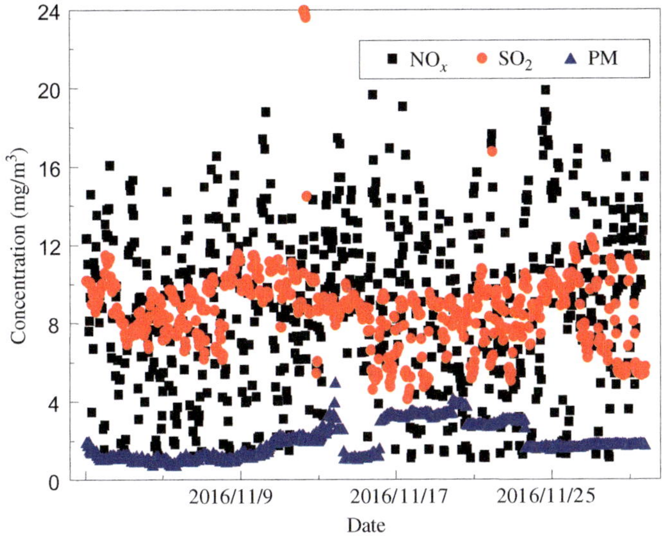

Fig. 4.6 Unit 1 emission data after ultra-low emission.

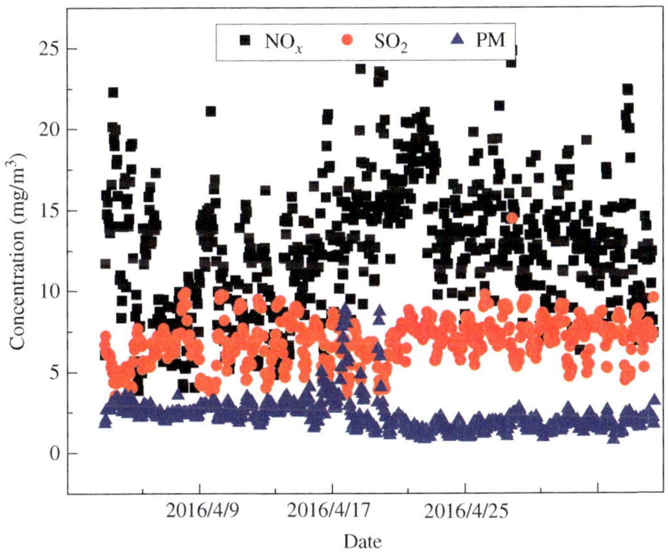

Fig. 4.7 Unit 2 emission data after ultra-low emission.

ultra-low emission in Wanzhou Power Plant. The emission concentrations of PM, SO_2, and NO_x were basically lower, at different times, than 5, 15, and 25 mg/m^3, which was much lower than the ultra-low emission standards of 10, 35, and 50 mg/m^3, respectively.

Table 4.4 Average emission concentration of pollutants in Wanzhou Power Plant

	NO_x (mg/m^3)		SO_2 (mg/m^3)		PM (mg/m^3)	
Time	**Unit 1**	**Unit 2**	**Unit 1**	**Unit 2**	**Unit 1**	**Unit 2**
January 2016	16.87	14.11	8.88	11.14	1.69	3.15
February 2016	15.92	9.49	9.38	8.5	3.47	1.91
March 2016	8.83	8.81	6.86	10.66	3.07	2.19
April 2016	–	12.05	–	11.46	–	2.18
May 2016	–	10.3	–	12.02	–	2.33
June 2016	16.54	–	11.34	–	3.31	–
July 2016	16.52	191.11	7.09	16.99	2.23	6.87
August 2016	29.62	26.09	10.05	10.22	2.96	1.64
September 2016	–	13.01	–	12.16	–	1.62
October 2016	17.86	24.82	5.16	8.1	2.2	2.51
November 2016	8.64	–	8.68	–	1.91	–
December 2016	6.71	11.52	8.59	8.22	2.46	1.81
January 2017	16.05	10.84	19.37	5.42	1.47	2.13

Note: The missing data in the table are caused by unit shutdown.

Table 4.4 illustrates the monthly average of emissions concentration of flue gas pollutants, originated from the two power units from January 2016 to January 2017. It shows that the annual average PM emission concentration was 2.48 mg/m^3 for Unit 1 and 2.58 mg/m^3 for Unit 2. The annual average of SO_2 emission concentration was 9.54 mg/m^3 for Unit 1 and 10.44 m/m^3 for Unit 2. The annual average of emission concentration of NO_x was 15.36 mg/m^3 for Unit 1 and 30.20 mg/m^3 for Unit 2. The average emission concentration of PM, SO_2, and NO_x from the two coal-fired units in Wanzhou Power Plant was lower than the emission concentration limit of atmospheric pollutant caused by gas turbines. It was in compliance with the ultra-low emission standards. And, the emission reduction effects were rather remarkable.

From the experience of Wangqu Power Plant and Wanzhou Power Plant, it could be seen that reconstruction and modernization of the traditional pollutant removal technology was an effective way to reduce emissions. New and up-to-date environmental protection requirements were fully considered for the then-to-be-built power units in Wanzhou Power Plant even during the project design stage. Accordingly, through optimization of project designs and the operational process, the ultra-low emission requirements were eventually reached. It indicated that improving index management under appropriate conditions would be conducive to strike a balance between investment and the technical requirement.

4.1.2 WESP

4.1.2.1 Zhoushan Power Plant

(1) Introduction to the power units and their ultra-low emissions implementation

Zhejiang Zhoushan Power Plant's power Unit 4 was a 350-MW supercritical coal-fired power unit made in China. The boiler was an SG-1146/25.4-M4409 type made by Shanghai Boiler Works Co., Ltd., with a design efficiency of 94%. On April 28, 2013, Zhejiang Zhoushan Power Plant made the decision to adopt WESP technology for Unit 4 in order to improve its PM emission standard and reduce the PM emission to 5 mg/m^3, aiming to meet the near-zero emission goal. In August 2013, Zhejiang Electric Power Design Institute was entrusted by Zhejiang Zhoushan Power Plant to develop a technical design of WESP. In June 2014, Zhejiang Zhoushan Power Plant's Unit 4 became the first newly built coal-fired power unit with near-zero emission that passed the environmental protection inspection successfully in China. The application of the first domestically made WESP in the near-zero emission exemplary project in coal-fired power plants was therefore coming into being. The overall technology roadmap includes low-NO_x burner + SCR + Electrostatic Precipitator (with 4 conventional electrodes + 1 rotating electrode) + seawater desulfurization + wet Electrostatic Precipitator (WESP).

A three-step technical measure was adopted to assure the near-zero PM emission. Firstly, the 5-field dry electrostatic precipitator (4 electric field + rotating electric field) was used for PM removal. Secondly, a portion of the PM was cleared by seawater desulfurization. And finally, the WESP was used for the most detailed treatment. Among these, high-frequency power technology was employed for electrostatic precipitators. The DESP's design efficiency was 99.94%, with the PM concentration of $\leq$30 mg/m^3 at the stack outlet.

The last electric field of ESP was of the rotating electrode type. Its advantages included the continuous clean condition of the electrode plate, prevention of back ionization, and assurance of precipitator's efficiency. Its advantages also included change of traditional rapping cleaning to PM brush cleaning, and placing of the PM cleaning brush at the non-PM collection area in order to minimize probability of secondary PM. The high-frequency power technology employed could increase the corona power when compared with the power frequency power supply and could increase the charging effect of the electric field PM. Furthermore, the average and current voltage were high, the device was small in size and light in weight, and the spark control characteristic of the high-frequency power source was good. Also, the output ripple was small and the average voltage of the electric field was high.

The WESP was equipped with the metal electrode plate, with the design PM removal efficiency of 70%, the design value of PM concentration of 16.5 mg/m^3 at the stack inlet, and PM concentration of $\leq$5 mg/m^3 at the stack outlet. The metal electrode plate used in this project was equipped with the horizontal flue gas WESP, which was again installed with a spray system. It was developed and manufactured by Zhejiang Feida Environmental Protection Technology Co., Ltd. Its electrode model came with the CN anode Plate and DS stitch lines.

In terms of desulfurization, Zhejiang Zhoushan Power Plant's power Unit 4 adopted seawater desulfurization technology. It was the first of its kind to have passed the inspection of the Environmental Protection Assessment Center of the Ministry of Environmental Protection of China, with a desulfurization efficiency of no less than 98%. During the near-zero emission implementation process, the flow rate of the spray water in the absorption system was able to be increased through optimal designing, and the flow rate of the flue gas was able to be slowed down through further expanding of the absorption system diameter, to increase the gas-liquid contact time. Simultaneously, the design of the aeration pipe network underwent a further optimizing process, so that the seawater desulfurization efficiency was elevated from 97% to 98%, and the SO_2 emission concentration decreased from 50 to 35 mg/m^3.

The whole package of technical solution for NO_x control included the application of low-NO_x burner technology and the installation of a well-configured SCR denitrification device, in order to make the value of the NO_x emission less than 50 mg/m^3 in coal-fired power plants.

Fig. 4.9 shows Unit 4 actual atmospheric pollutants emission data. It indicates that the concentration of particles was generally between 2 and 3 mg/m^3, and the concentration of SO_2 around 5 mg/m^3 under different load conditions. Through comparison with the power units later built in other plants, it was found that the SO_2 concentration value became lower after the seawater desulfurization treatment, which meant that this kind of desulfurization effect was obviously better than the limestone gypsum method. The NO_x concentration was normally between 15 and 40 mg/m^3. The flow chart for Zhoushan Power Plant is as follows (Fig. 4.8).

Fig. 4.8 Roadmap of Zhoushan Power Plant.

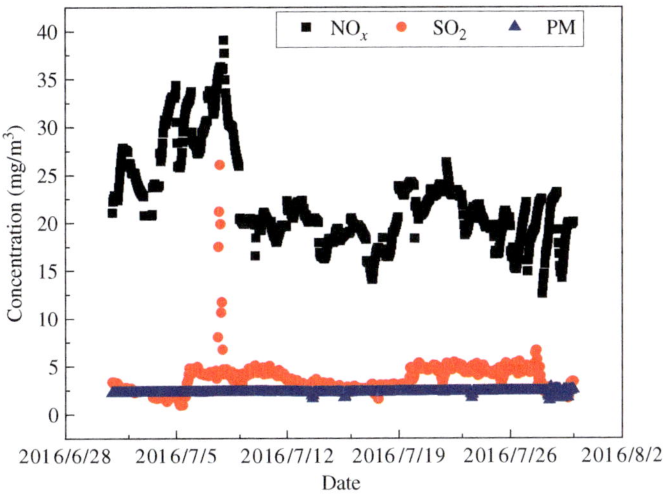

Fig. 4.9 Unit 4 emission data after ultra-low emission.

4.2 Emissions control technology based on LTT-ESP

4.2.1 Yuanyanghu Power Plant

(1) Introduction to the power units before upgrade

Yuanyanghu Power Plant was equipped with 2 × 660 MW supercritical coal-fired air-cooled power units. The boiler was a 2141 t/h supercritical once-through boiler produced by Shanghai Boiler Works Co., Ltd. The steam turbine was a supercritical, one intermediate reheater, three cylinders, four exhaust hoods, and direct air-cooled condensing steam turbine, produced by Dongfang Steam Turbine Co., Ltd. The generator was a three-phase two-pole synchronous generator produced by Dongfang Electric Machinery Co., Ltd.

PM removal system

Each furnace was equipped with four electric field electrostatic precipitator equipment, accompanied by online soot blowing devices. The design concentration was 20 mg/m^3 at the electrostatic precipitator outlet, and ESP's actual discharge concentration was 18.42 mg/m^3.

Desulfurization system

This system was equipped with two sets of wet limestone gypsum flue gas desulfurization equipment, installed at the rear part and the side of the boiler stack respectively, with an annual operation of 100%. The design sulfur content in coal was 0.93% with a desulfurization efficiency of more than 98%. Under the power unit's full load condition, the SO_2 concentration in flue gas at the inlet of WFGD was about 2100 mg/m^3, and the SO_2 emission concentration was less than 35 mg/m^3.

Denitrification system

In terms of denitrification, the two power units adopted a selective catalytic reduction (SCR) denitrification technique and used the liquid ammonia as reducing agent. Each boiler was equipped with two SCR reactors, which was placed between the economizer and the air preheater, arranged symmetrically along the center line of the boiler.

(2) Introduction to the upgrade roadmap

A low-temperature economizer was installed before the inlet of electrostatic precipitator, to reduce ESP's flue gas temperature to 95°C at its inlet. A three-phase high-frequency power supply was installed as an alternative, while the distance between ESP's electrode plate and the electrode line was adjusted accordingly. Through these methods, the PM removal efficiency was improved. And, as to the two power units, the PM concentration at ESP outlet remained at a steady level, and was controlled to be less than 20 mg/m^3.

For these two powers units desulfurization upgrade, the double-pallet technology under US B&W patent was adopted. Besides, two layers of alloy trays were installed inside the absorption tower, and three spray layers were changed (the first layer was changed into a cross-spray type). Upgrades were performed on the slurry circulation pump and the dehydration system. Thus, the desulfurization efficiency became larger than 98.78%, to ensure that the SO_2 emission concentration was at least 35 mg/m^3. In Unit 1 desulfurization upgrade, an additional layer of tubular demisters was installed on the basis of the originally installed two-layer roof ridge demister. In Unit 2 desulfurization upgrade, a highly efficient demister was installed with the sufficient consideration to reduce the fog droplets carried with the flue gas.

The denitrification efficiency improvement for the two units was completed successfully in December 2012. Based on the operational characteristics of the current denitrification equipment, and through adjustments applied to the burner, NO_x concentration at the inlet of denitrification units was reduced. And adjustment was also applied to the system's ammonia injection amount. All these measures were taken in order to ensure that NO_x emission concentration would be under control and stay below the standard value at the outlets of the two denitrification systems. The flow chart about Yuanyanghu Power Plant before and after upgrade is as follows (Fig. 4.10).

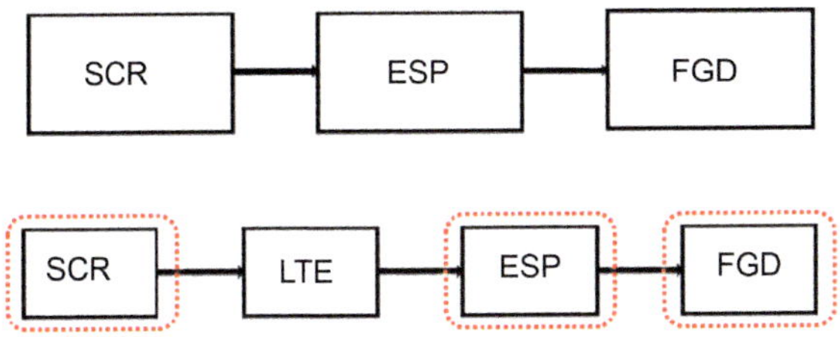

Fig. 4.10 Upgrade roadmap of Yuanyanghu Power Plant.

Table 4.5 Analysis on coal quality in Yuanyanghu Power Plant

Time	Coal type	Moisture (%)	Ash (%)	Sulfur (%)	Volatile matter (%)	Net calorific value (MJ/Kg)
2014	Bituminous	12.32	24.21	0.85	22.12	20.27
2015	Bituminous	13.09	23.15	0.94	21.98	19.13
2016	Bituminous	11.61	24.49	0.76	23.49	19.22
Mean	Bituminous	12.34	23.95	0.85	22.53	19.54

(3) Coal quality

Both the design coal and alternative checking coal used by power units were supplied by Ningdong Coalfield & Coal Mine. Tests were made to the quality of coal used in May 2014, November 2014, May 2015, November 2015, May 2016, and October 2016. The specific coal quality was tabulated based on the date, as shown in Table 4.5. And the quality of coal used was analyzed in the following figures based on the date too.

It can be seen from the figure that the average value of total moisture content on the as-received basis was 12.34%, with a limited moving range. And the average value of the ash content on as-received basis is 23.95%, while the moving range was so limited that its influence could be ignored. The average value of sulfur content on as-received basis was 0.85%, remaining almost at a stable level; the average value of the volatile matter content was 35.36%, generally remaining at a constant level without much fluctuation. On the whole, the coal quality did not show much fluctuation before and after the ultra-low emission upgrade (Fig. 4.11).

(4) Analysis on the operational characteristics of ultra-low emission

Figs. 4.12 and 4.13 show the emission characteristics of PM, SO_2, and NO_x in flue gas at different times before and after the ultra-low upgrade for Unit 1 in Yuanyanghu Power Plant. The NO_x emission concentration of Unit 1 mostly fluctuated between 50 and 70 mg/m^3 before the ultra-low emission upgrade. Not a small number of data showed far exceeding 100.00 mg/m^3, while the peak value even reached as high

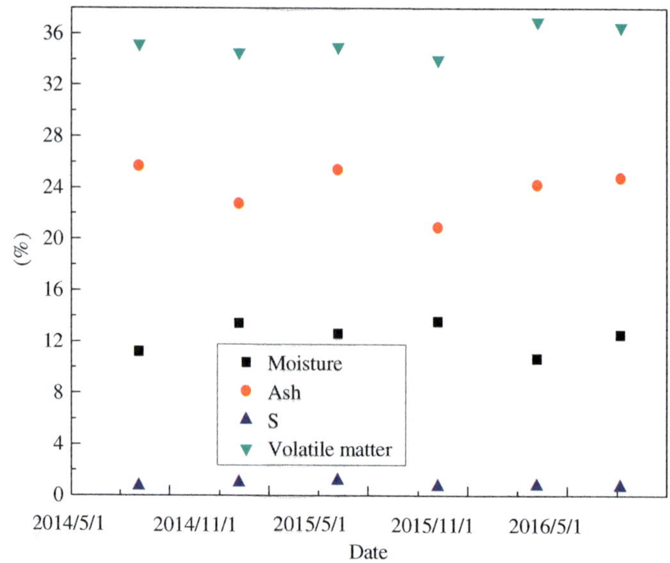

Fig. 4.11 Coal quality in Yuanyanghu Power Plant.

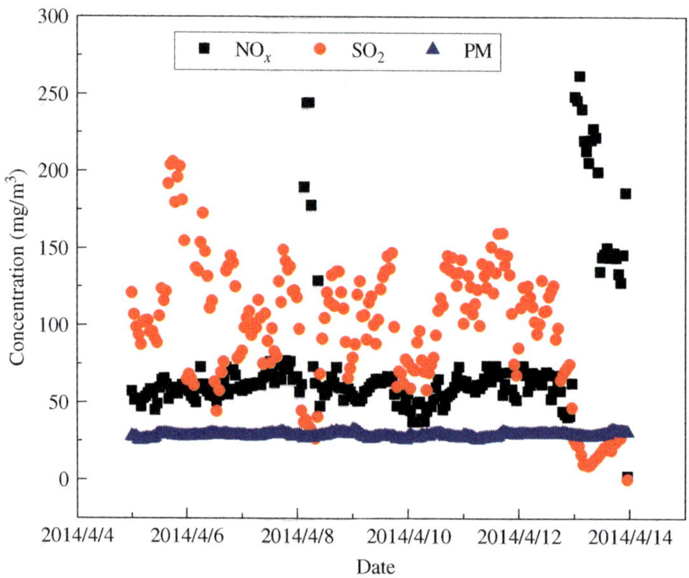

Fig. 4.12 Emission data before ultra-low emission upgrade for Unit 1.

as 262.10 mg/m^3, indicating a rather serious emission scenario. Still, SO_2 concentration fluctuated mostly between 75 and 125 mg/m^3, but with a quite significant moving range. The PM concentration though fluctuated within the range of 25 ~ 30 mg/m^3, but was remarkable higher

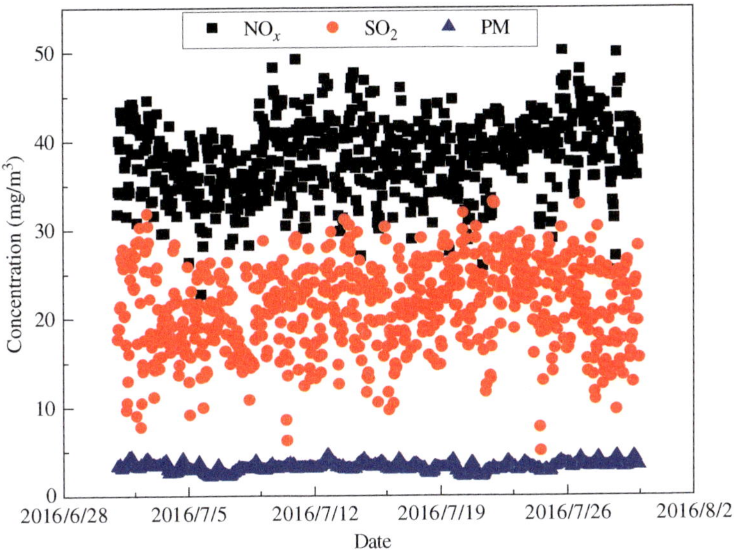

Fig. 4.13 Emission data after ultra-low emission upgrade for Unit 1.

than the ultra-low emission standard. However, after the ultra-low emission upgrade, NO_x emission concentration of Unit 1 saw a significant cutback, ranging mostly between 35 and 45 mg/m^3. And even the peak value read only 50.17 mg/m^3. The numbers were on the whole lower than the ultra-low emission standard. Furthermore, the SO_2 emission concentration showed an apparent reduction, with emission concentration mostly running between 10 and 30 mg/m^3. Its maximum value was 33.11 mg/m^3, less than the ultra-low emission standard of 35 mg/m^3. And, the PM emission concentration was less than 5 mg/m^3, which could remain at a stable level and could meet the ultra-low emission requirements.

Figs. 4.14 and 4.15 show the emission characteristics of PM, SO_2, and NO_x in flue gas at different times before and after the ultra-low upgrade for Unit 2. The NO_x emission concentration of Unit 2 mostly fluctuated between 50 and 75 mg/m^3 before the ultra-low emission upgrade. Some data even showed far exceeding 150.00 mg/m^3. Still, SO_2 concentration fluctuated mostly between 50 and 150 mg/m^3. PM concentration fluctuated within the range of 20~30 mg/m^3. All these readings failed to be in compliance with the ultra low emission standard. However, after ultra-low emission upgrade, NO_x emission concentration of Unit 2 saw a significant reduction, mostly remaining between 35 and 50 mg/m^3. Though the peak value did read 50.94 mg/m^3, it was quite close to the ultra-low emission standard of

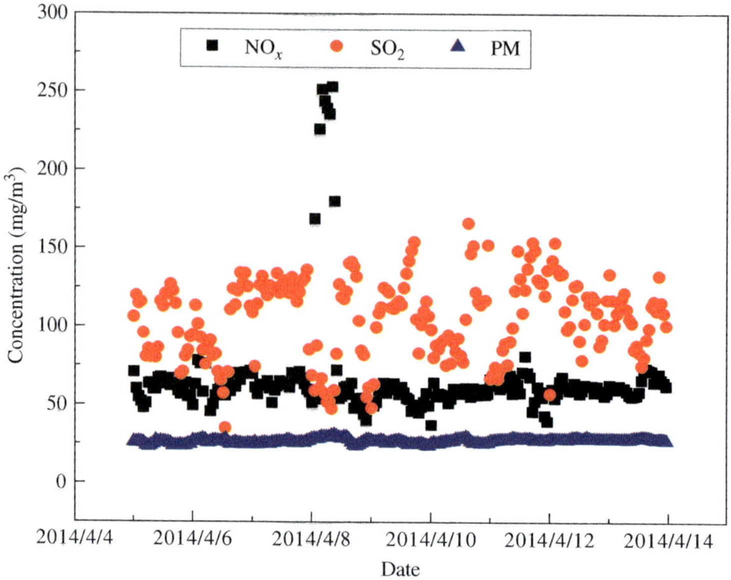

Fig. 4.14 Emission data before ultra-low emission upgrade for Unit 2.

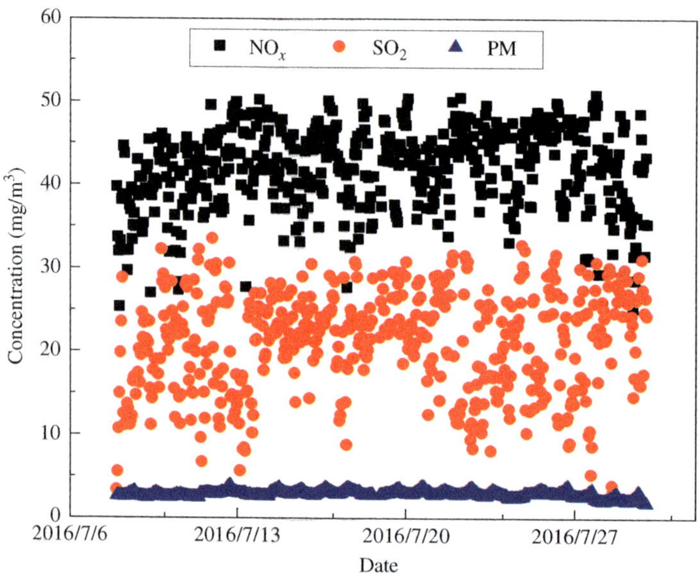

Fig. 4.15 Emission data after ultra-low emission upgrade for Unit 2.

50 mg/m^3. Furthermore, the SO_2 emission concentration showed an apparent reduction, with emission concentration mostly remaining between 10 and 30 mg/m^3. It had a maximum value of a mere 33.56 mg/m^3, still lower than the ultra-low emission standard of 35 mg/m^3. And, PM emission concentration was less than 5 mg/m^3, which could meet the ultra-low emission requirements on a steady basis.

4.2.2 Hequ Power Plant

(1) Introduction of the power units before upgrade

The Hequ Power Plant was equipped with 2×600 MW coal-fired supercritical air-cooled condensers units, designated as Units 3 and 4. The boiler was a supercritical control circulating drum boiler, manufactured by Harbin Boiler Factory Co., Ltd. The steam turbine was a supercritical, one intermediate reheater, single-shaft, three-cylinder, four exhaust hoods, and direct air-cooled condensing steam turbine, manufactured by Dongfang Steam Turbine Co., Ltd. The electric generator was of a three-phase two-pole synchronous type, manufactured by Dongfang Electric Machinery Co., Ltd.

PM removal system

Each boiler was equipped with 1 set of fabric filter equipment, which was an external filter type for PM removal. It could allow online PM collection, online maintenance, and online replacement of the fabric filter's main body. It was also equipped with a bypass flue. A spraying system as designed was installed at its outlet. The design PM concentration was 33.8 g/m^3 at the fabric filter inlet, and 20 mg/m^3 at the fabric filter outlet. In any case, the emission concentration should not exceed 30 mg/m^3 for its design. The actual emission concentration was 20.42 mg/m^3.

Desulfurization system

The boiler in Hequ Power Plant was equipped with two sets of wet limestone gypsum flue gas desulfurization devices. Based on the plant's general layout planning, desulfurization stations were designed and constructed at the same pace as the on-going construction of the main project. The desulfurization equipment was positioned at the rear and side portion of the boiler stack. Currently, the overall operational condition of the desulfurization equipment was quite satisfactory, with an annual operating rate of 100%. It reached a desulfurization efficiency of over 95%, with a 0.73% sulfur content in coal. Before the ultra-low emission upgrade, when power units were operating under continuous full load condition, SO_2 concentration in flue gas at the stack inlet was 3840 mg/m^3, failing to meet the requirement of SO_2 emission concentration of less than 35 mg/m^3.

Denitrification system

This system adopted selective catalytic reduction (SCR) denitrification techniques. And liquid ammonia–producing techniques were adopted as a reducing agent. Each boiler was equipped with two SCR reactors, arranged symmetrically along the center line of the boiler. SCR reactors were positioned between the economizer and the air preheater.

(2) Introduction of upgrade roadmaps

For Units 3 and 4, the original fabric filters were replaced with electrostatic precipitators. A low-temperature economizer was positioned at the front portion of the precipitator, which means that a technology roadmap of LTT-ESP was adopted. Flue and PM collector chambers were additionally installed at the two sides of the same place occupied previously by the removed filter chamber. A new electric field also was added, and finally making the system become a double-chamber, ten-channel and five-field electric field. New anode and cathode systems were installed too. Thus, after the upgrade, the electrostatic precipitator's first and second electric fields were powered by small partitions, while third, fourth, and fifth by regular partitions. Totally, 38 sets of ZH series high-efficiency three-phase power were installed.

For wet limestone gypsum flue gas desulfurization device, spin-coupled devices were installed and auxiliary spray desulfurization technology was used, in which two auxiliary spray layers were added for the purpose of sidewalls escape prevention.

The traditional low-efficiency two-layer roof ridge style demister as originally installed was removed, and a high-efficiency three-layer roof style demister, which had been newly improved, was adopted in this upgrade. The goal was to increase demister efficiency and contribute to collaborate PM removal efforts.

The two layers previously SCR catalyst equipped Sides A and B of flue were removed and replaced with alternatives. Among them, Side A was replaced with a three-layer conventional catalyst, and Side B a three-layer nontoxic catalyst. An acoustic sonic soot-blowing system was also added as part of the whole configuration. Also, improvement was executed for flow equalization in the SCR flue. The flow chart about Hequ Power Plant before and after the upgrade is as follows (Fig. 4.16).

(3) Coal quality

The coal used in the coal-fired power units in Hequ Power Plant was lignite type, supplied by Yuquan Coal Mine and Huangbai Coal Mine. Quality tests of the coal used in this plant during the whole year of 2014, 2015, and 2016 were conducted. Specific coal quality is shown in Table 4.6.

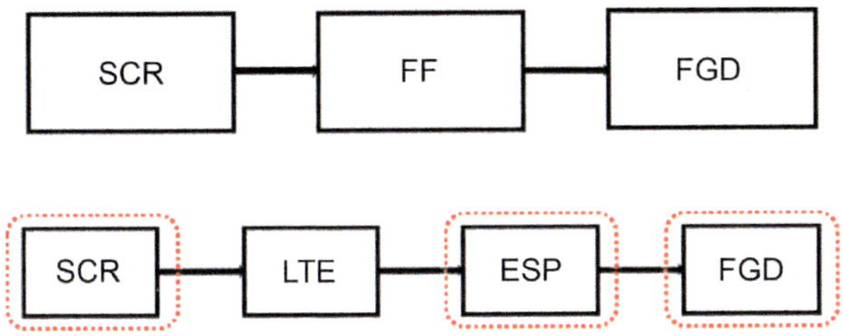

Fig. 4.16 Upgrade roadmap of Hequ Power Plant.

Table 4.6 Analysis on coal quality in Hequ Power Plant

Time	Coal type	Moisture (%)	Ash (%)	Sulfur (%)	Volatile matter (%)	Low calorific value (MJ/Kg)
2014	Lignite	11.53	20.7	0.61	27.15	17.97
2015	Lignite	10.7	21.61	0.55	26.94	18.11
2016	Lignite	9.43	18.33	0.55	23.72	16.63
Mean	Lignite	10.55	20.21	0.57	25.94	17.56

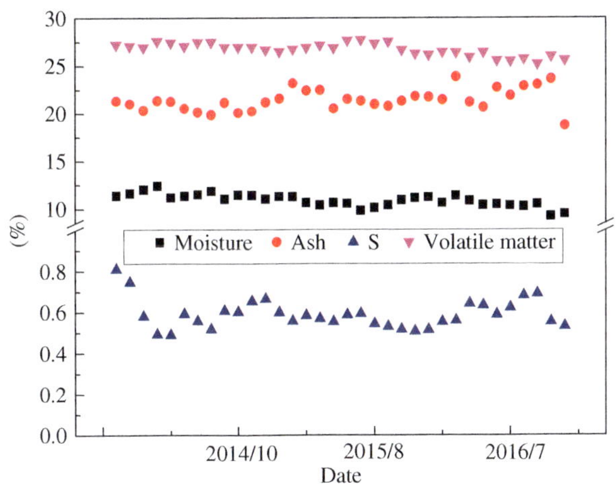

Fig. 4.17 Analysis on coal quality in Hequ Power Plant.

Fig. 4.17 shows the analysis on the quality of coal used in coal-fired units of Hequ Power Plant. Average value of total moisture content on the received basis was 10.55%, with the moving range basically within 1%. During the three-year test, average value of the ash content on as-received basis was 20.21%, with a moving range of around 2%, and a slight increase in the later period. The average value of the sulfur content on as-received basis was 0.57%, which showed a dramatic fluctuating trend, with a peak value reaching over 0.8% and a fluctuation range of 0.4%–0.8%. Average value of the volatiles content on as-received basis was 25.94%, with a fluctuation of less than 1% and a slight decrease in the later period.

(4) Analysis on results of ultra-low emission upgrade

Figs. 4.18 to 4.21 show the emission characteristics of PM, SO_2, and NO_x in flue gas before and after the ultra-low emission upgrade to the two power units in Hequ Power Plant. It can be seen from the Figures that emission concentration of NO_x remained at 60–85 mg/m^3 before the power units were upgraded. SO_2 concentration of Unit 3 mostly remained between 60 and 90 mg/m^3, while that of Unit 4 was between 40 and 90 mg/m^3. PM concentration was basically in the range of 5–10 mg/m^3.

In contrast, after the ultra-low emission upgrade, the three pollutants emission concentration for Unit 3 and 4 decreased significantly. NO_x concentration was generally less than 45 mg/m^3, SO_2 concentration was lowered to 10–30 mg/m^3, and PM concentration dropped below 5 mg/m^3. It was obvious that this upgrade yielded a quite satisfactory result.

Table 4.7 offers the data on Unit 4 of Hequ Power Plant, collected from factory's acceptance test of the upgraded power units in operation, within one year after Unit 4's completion of the ultra-low emission upgrade. It can be seen that, after its upgrade, the average emission

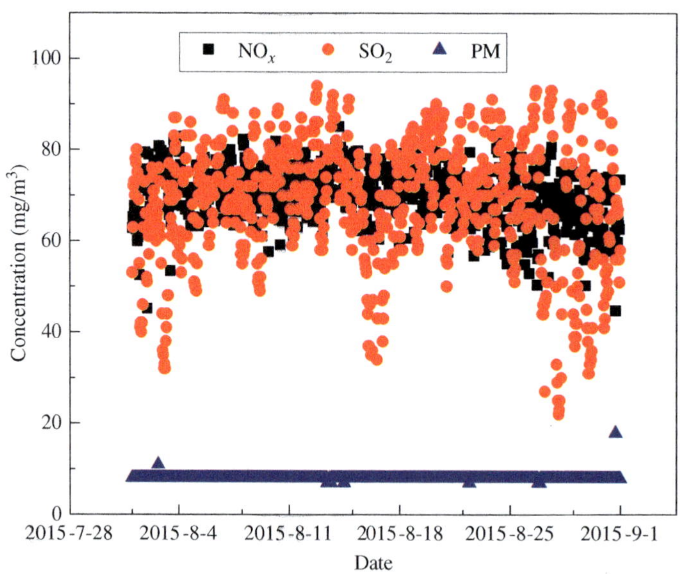

Fig. 4.18 Emission before ultra-low emission upgrade for Unit 3.

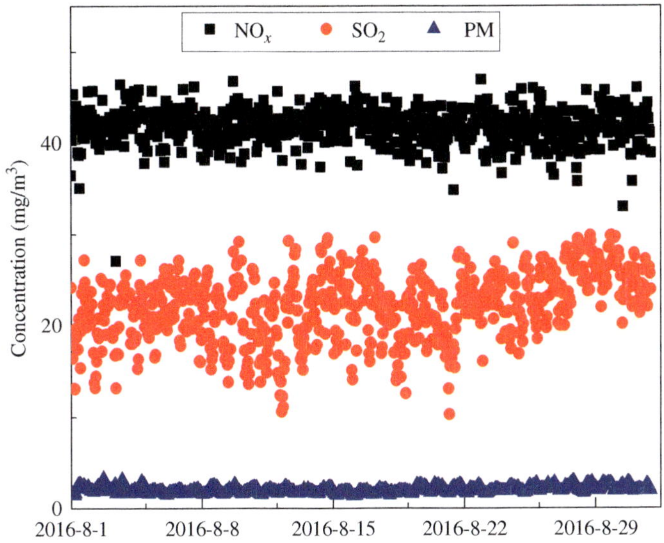

Fig. 4.19 Emission after ultra-low emission upgrade for Unit 3.

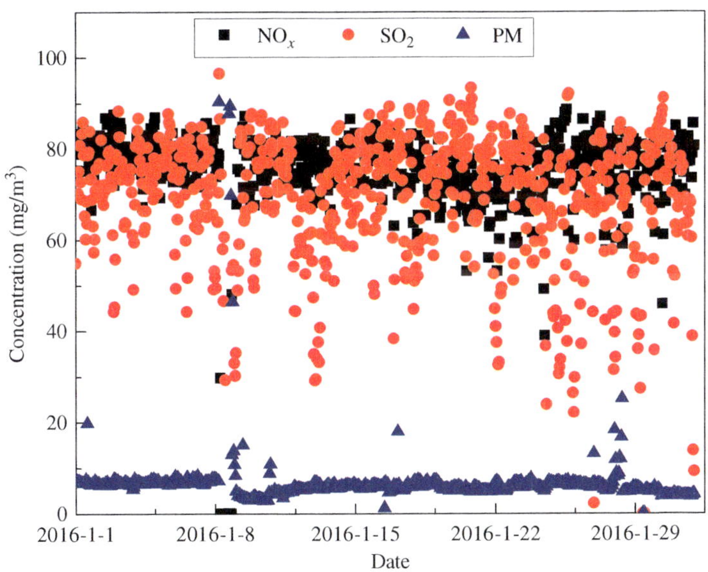

Fig. 4.20 Emission before ultra-low emission upgrade for Unit 4.

concentration of PM, SO_2, and NO_x was lower than emission concentration limit of atmospheric pollutants caused by gas turbines. It was in compliance with the ultra-low emission standards, and possessed a remarkable upgrade and emission reduction effects.

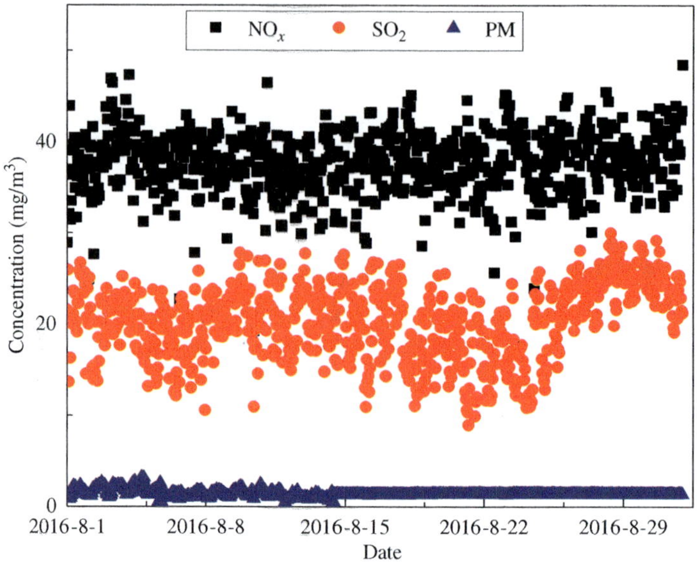

Fig. 4.21 Emission after ultra-low emission upgrade for Unit 4.

Table 4.7 Average emission data after ultra-low emission upgrade for Unit 4

	Monthly	Minimum daily	Maximum daily
PM	1.98	0.85	3.52
SO_2	21.3	15.2	27.0
NO_x	40.4	16.8	44.2

4.2.3 Dagang Power plant

(1) Introduction of power unit before upgrade

Dagang Power Plant were equipped with four 328.5-MW coal-fired power units, with a total installed capacity of 1,314 MW. Among them, Units 1 and 2 were put into operation for electric generation in 1978 and 1979, respectively, and Units 3 and 4 in 1991 and 1992, respectively.

PM removal system

The four coal-fired power units were equipped with electrostatic precipitators which all belonged to horizontal DESP. Before the upgrade, PM concentration at the stack outlet was around 30~50 mg/m^3.

Desulfurization system

When Units 1 and Unit 2 were undergoing the upgrade from oil-fired to coal-fired in 2002, the limestone wet flue gas desulfurization device was installed. Due to variations of the coal quality, it could not meet the national SO_2 emission standard. In August 2008, a flue gas desulfurization upgrade project was performed on Units 3 and 4, and a wet limestone gypsum flue gas desulfurization technique was implemented.

Denitrification system

Before the ultra-low emission upgrade, NO_x removal was mainly dependent upon low-nitrogen combustion inside the furnace, such as depending on low-nitrogen burners to bring down NO_x formation, but without any denitrification treatment of the flue gas. As a result, NO_x concentration in the flue gas could not meet the ultra-low emission requirements.

(2) Introduction of upgrade roadmap

In view of the main problems existing in the previous PM removal, Dagang Power Plant decided to select an upgrade plan, comprising an ESP with three-phase high-efficiency power supply upgrade + low temperature economizer (LTE) upgrade, both featuring a low investment budget, a short construction period, and a small project amount. Through installation of a low-temperature economizer at the front of the ESP inlet for all of the four units, the flue gas temperature at their inlet was reduced, and the PM removal efficiency of the ESP was greatly improved. Thus, PM concentration at the stack outlet was kept stably below 10 mg/m^3.

In June 2010, an upgrade was performed on Unit 1 and 2 desulfurization equipment. In order to reach the mandatory standard required for demonstration power plants, a kind of collaborate PM collection means was adopted for Units 3 and 4. Furthermore, the demisters of four units limestone wet desulfurization absorption system not only underwent an upgrade process, but also were replaced with a high-efficiency demister of a three-stage roof ridge type.

Construction of the flue gas denitrification (SCR) system for the four power units was completed in 2013 and 2014. And, low NO_x burners and air preheaters were replaced with alternatives simultaneously. Furthermore, a "three-in-one" fan system upgrade was executed. These measures aimed at minimizing NO_x emission concentration, and stably keeping NO_x emission concentration at four denitrification devices outlets at a level below 50 mg/m^3. Thus, the ultra-low emission requirement could eventually be met. The flow chart about Dagang Power Plant before and after the upgrade is as follows (Fig. 4.22).

(3) Coal quality

The coal used by four coal-fired power units was all supplied by Shenhua Coal. Tests were done for the quality of coal admitted from January 2014 to September 2016. Specific coal quality was tabulated on a yearly basis, as shown in Table 4.8.

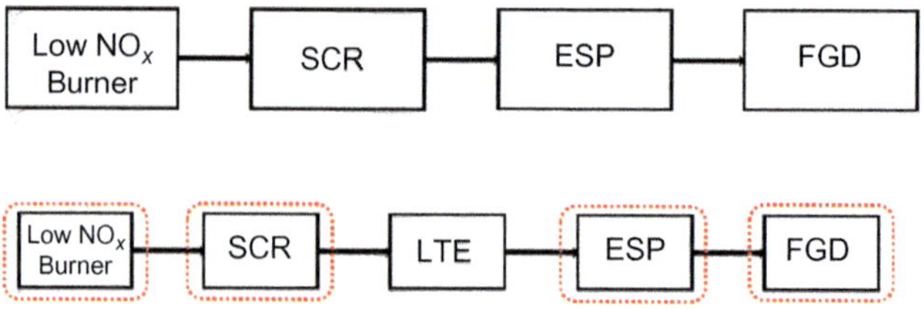

Fig. 4.22 Upgrade roadmap of Dagang Power Plant.

Table 4.8 Analysis on coal quality in Dagang Power Plant

Time	Coal type	Moisture (%)	Ash (%)	Sulfur (%)	Volatile matter (%)	Low calorific value (MJ/Kg)
2014	Bituminous	12.30	18.09	0.45	25.87	20.56
2015	Bituminous	12.97	16.98	0.42	26.23	20.69
2016	Bituminous	13.33	16.26	0.41	26.65	21.19
Mean	Bituminous	12.87	17.11	0.43	26.25	20.81

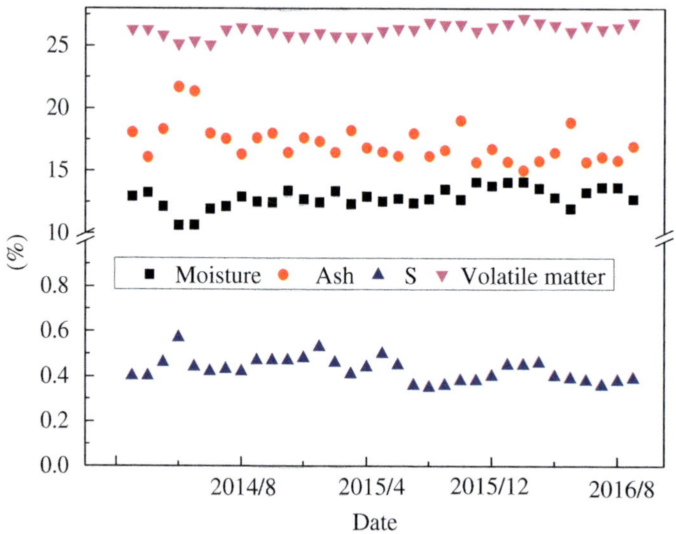

Fig. 4.23 Coal quality analysis in Dagang Power Plant.

According to analysis on coal quality from January 2014 to September 2016 as shown in the month-on-month Fig. 4.23, it can be seen clearly that, during a considerably long period before and after the integrated ultra-low emission upgrade, the quality of coal used by coal-fired units in Dagang Power Plant was quite stable. The average value of the total moisture content on the as-received basis was about 12.87%, with a moving range of within 2%. The average value of the ash content on as-received basis was 17.11%, with a slightly stronger fluctuation. Its moving range, except for April and May 2014, was all within 1%. The average value of the sulfur content on as-received basis was 0.43%, with a fluctuation range of basically within 0.1%. The average value of the volatiles content on as-received basis was 26.25%, whose moving range was less than 1%.

(4) Analysis on results of ultra-low emission upgrade

Figs. 4.24 to 4.28 show the emission characteristics of PM, SO_2, and NO_x for Units 1 and 2 before and after the ultra-low emission upgrade in Dagang Power Plant. It can be seen from these Figures that emission concentration of NO_x remained at 25~50 mg/m^3, SO_2 remained at 10~30 mg/m^3, and PM remained at 5~10 mg/m^3 before power units were upgraded. After the ultra-low emission upgrade, NO_x emission was in the range of 20~40 mg/m^3, SO_2 emission was without obvious change, and PM emission was less than 5 mg/m^3. Figs. 4.28 and 4.29 show the emission characteristics of PM, SO_2, and NO_x for Unit 3 before and after the ultra-low emission upgrade. Before the upgrade, the emission concentration of NO_x remained at 40~80 mg/m^3, peak value even was more than 90 mg/m^3. In most cases, SO_2 remained at

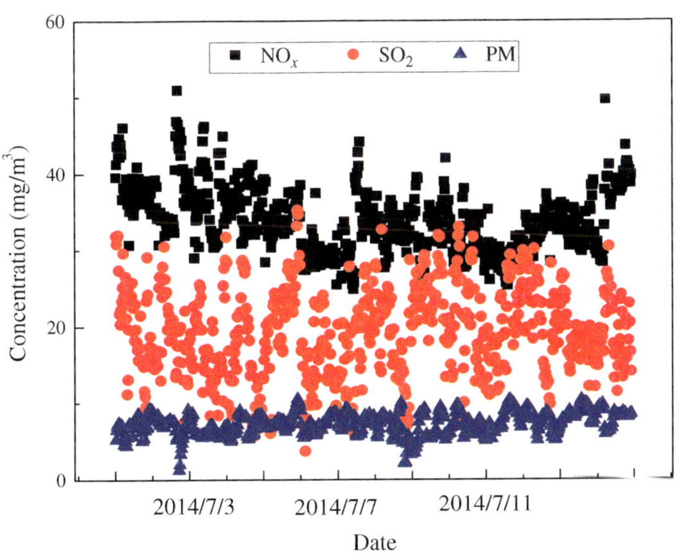

Fig. 4.24 Emission before ultra-low emission upgraded of Unit 1.

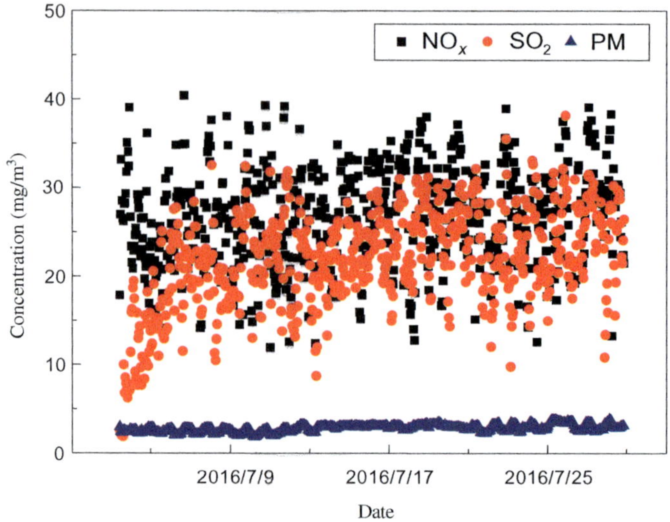

Fig. 4.25 Emission after ultra-low emission upgraded of Unit 1.

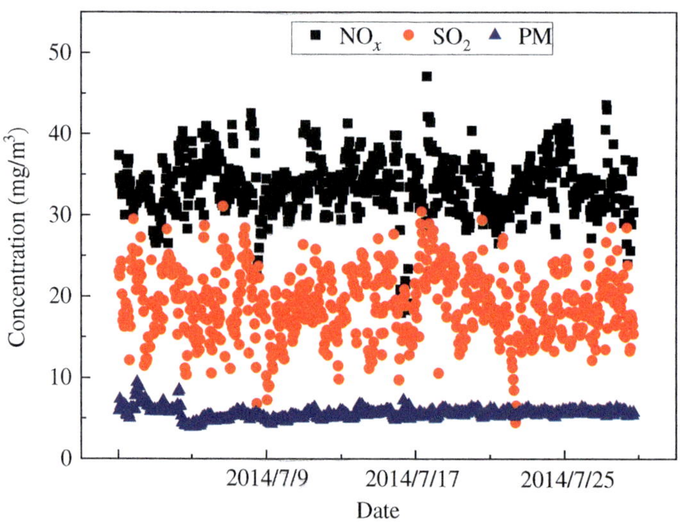

Fig. 4.26 Emission before ultra-low emission upgraded of Unit 2.

5~30 mg/m^3 and PM remained at 5~10 mg/m^3. After the ultra-low emission upgrade, NO_x emission was in the range of 20~40 mg/m^3, SO_2 emission was in the range of 5~25 mg/m^3, and PM emission was less than 5 mg/m^3.

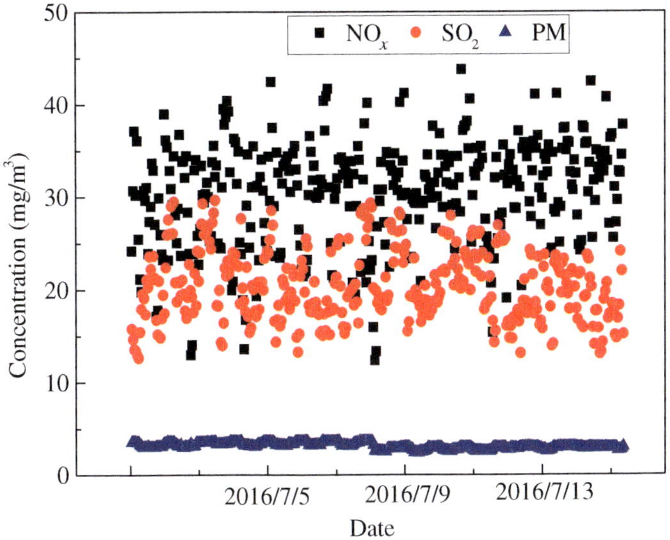

Fig. 4.27 Emission after ultra-low emission upgraded of Unit 2.

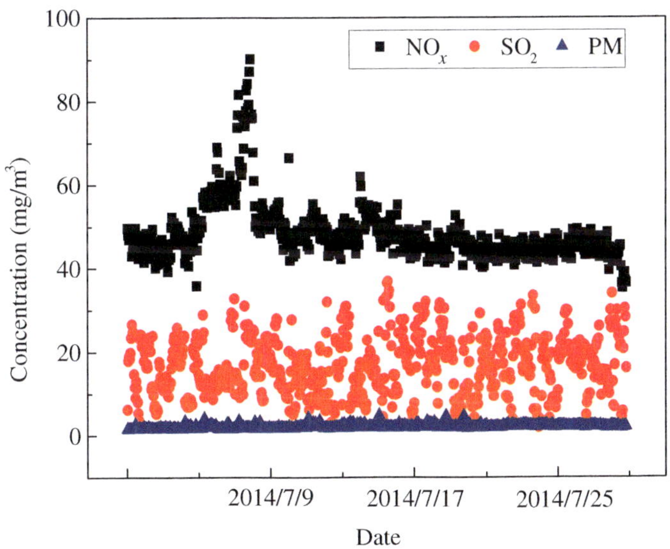

Fig. 4.28 Emission before ultra-low emission upgraded of Unit 3.

Figs. 4.30 and 4.31 show the emission characteristics of PM, SO_2, and NO_x for Unit 4 before and after the ultra-low emission upgrade. Before the upgrade, the emission concentration of NO_x remained at $20\sim40$ mg/m^3, SO_2 remained at $10\sim30$ mg/m^3. Some PM

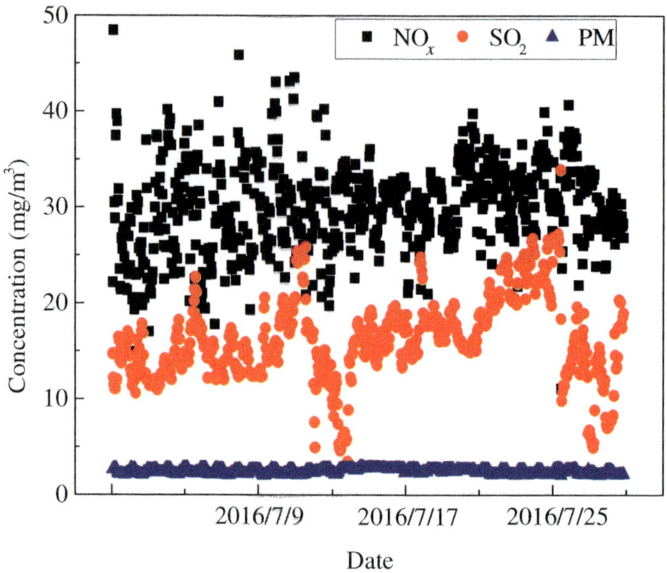

Fig. 4.29 Emission after ultra-low emission upgraded of Unit 3.

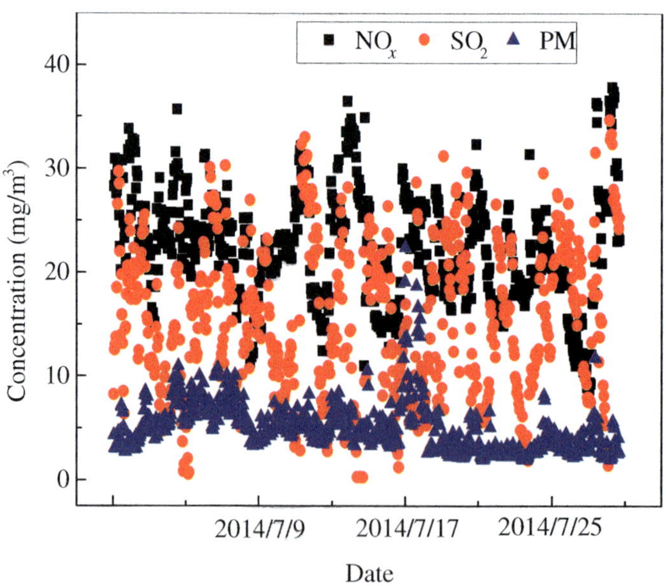

Fig. 4.30 Emission before ultra-low emission upgraded of Unit 4.

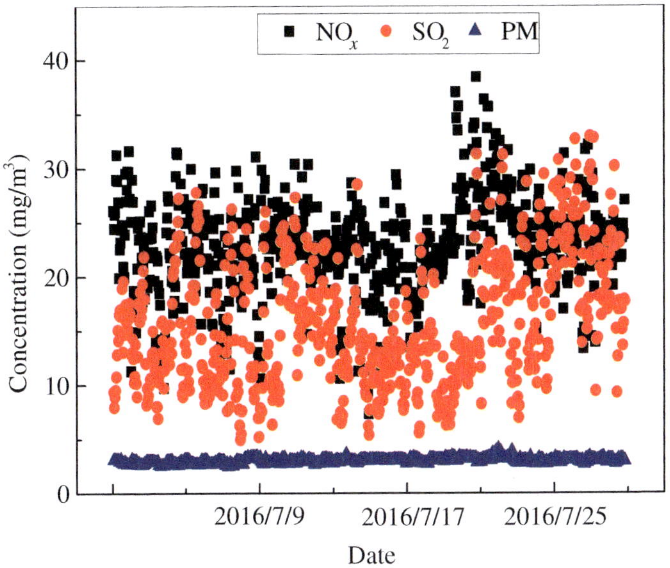

Fig. 4.31 Emission after ultra-low emission upgraded of Unit 4.

concentration exceeded 5 mg/m^{3}and the peak value was even more than 20 mg/m^3. After the ultra-low emission upgrade, NO_x emission was in the range of 20～30 mg/m^3, SO_2 emission was without obvious change, and PM emission was less than 5 mg/m^3.

From the running of Yuanyanghu Power Plant, Hequ Power Plant, and Dagang Power Plant, it's found that installation of the low-temperature economizer at the front of ESP could significantly decrease the PM emission and resulted in satisfactory emission reduction effects. Which kind of technology will be chosen is the results of the original pollutant emission level, cost and social conditions, etc. Due to the fact that the SO_2 emission concentration of Yuanyagnhu Power Plant used to be at a quite high level, double-pallet technology was adopted for the desulfurization upgrade. Dagang Power Plant is located in the Beijing-Tianjin-Hebei region, where a more stringent requirement on environmental protection had already been in practice. Thus, although this power plant had a comparatively low emission footprint, further in-depth emission cutback was still in urgent need.

From these upgrade cases, it could be seen that both the installation of low-temperature economizers at the front of ESP and the addition of WESP contributed greatly to obtaining good results on ultra-low emission. Thus, many power plants adopted both technologies simultaneously during their ultra-low emission upgrade process.

4.2.4 Sanhe Power Plant

(1) Introduction of power unit before upgrade

Sanhe Power Plant was constructed through two stages, namely, project phase I and project phase II. In phase I, two 350-MW condensing steam units, designated as Units 1 and 2, were imported from Mitsubishi of Japan. Its boiler was MB-FRR1175/20 subcritical, controlled circulation, solid-state slagging, and coal-fired steam drum furnace, manufactured by Mitsubishi Heavy Industries Kobe Shipyard Japan. Two units were put into operation for electric power generation in 1999 and 2000, respectively. And desulfurization upgrade was conducted in 2006.

In the project phase II, two 300-MW units were combined heat and power production power units, designated as Units 3 and 4. The boiler was a DG1025/18.2-116 subcritical, natural circulation, and solid-state slag-discharging furnace, manufactured by Dongfang Boiler (Group) Co., Ltd. Desulfurization upgrade was conducted on both units simultaneously. And they were put into operation in August and November 2007, respectively.

(2) Introduction of upgrade roadmap

In the year 2013, Unit 1 of Sanhe Power Plant achieved ultra-low emission successfully, and became the first upgraded ultra-low emission unit in the whole Beijing-Tianjin-Hebei region. Up to November 2015, all four units of Sanhe Power Plant achieved ultra-low emission successfully.

The ultra-low emission upgrade included the low-nitrogen burner upgrade, the addition of low-temperature economizers, replacement of the power-frequency power supply to high-frequency ones, desulfurization efficiency improvement (In Phase I, a spray layer and slurry circulation pump were added to the desulfurization absorption system, two-layer plated demisters were replaced with two-layer ridge demisters + tube demisters, and oxidation fans were replaced. And in Phase II, a spin-coupled device was added into the desulfurization absorption system, two-layer plated demisters were replaced with tube bundle-type ESP, and oxidation fans were replaced), and additional installation of WESP (except for Unit 3).

The flow chart about Sanhe Power Plant before and after its update is as follows (Fig. 4.32).

The coal used by four power units of Sanhe Power Plant all belonged to the same quality. WESP of Unit 1 was different from that of Units 2 and 4 in the aspect that the former used a flexible electrode plate WESP, while the

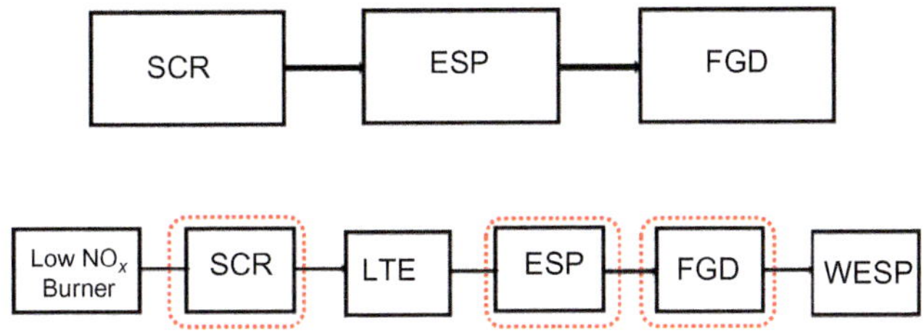

Fig. 4.32 Sanhe Power Plant's upgrade roadmap.

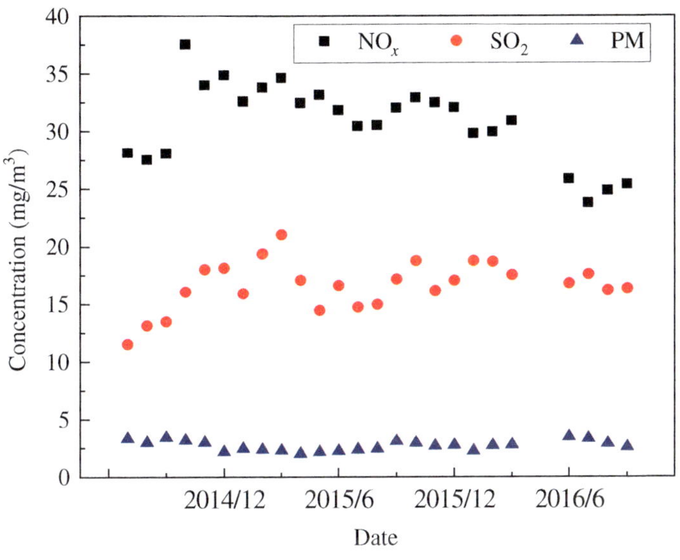

Fig. 4.33 Unit 1 emission status after ultra-low emission upgrade.

latter used a rigid electrode plate one. For Unit 3, there was no WESP equipped.

Fig. 4.33 shows the emission characteristics of PM, SO_2, and NO_x emissions of each unit under different times after the ultra-low emission upgrade Hebei Sanhe Power Plant. It's obvious from the figure that the characteristics of PM, SO_2, and NO_x emissions were generally lower than 5, 20, 40 mg/m^3, respectively (Figs. 4.34–4.36).

4.2.5 A power plant in Zhejiang province

(1) Introduction of units before upgrade

This power plant is located in eastern coastal area of China. Its 2 × 1000 MW power units (Units 7 and 8) were of ultra-supercritical

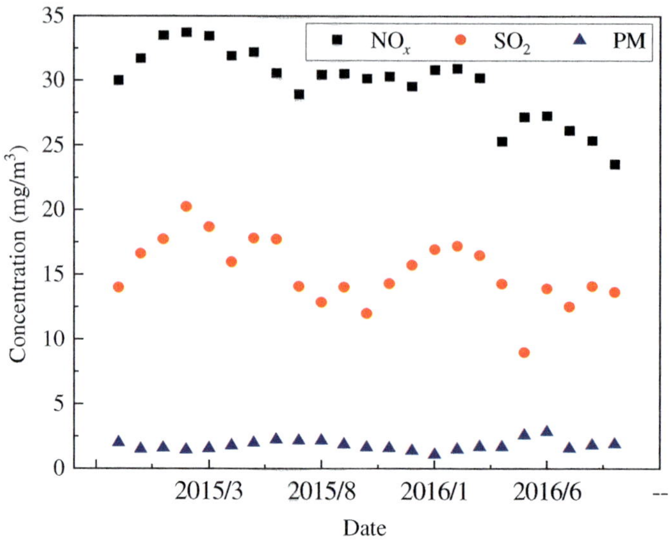

Fig. 4.34 Unit 2 emission status after ultra-low emission upgrade.

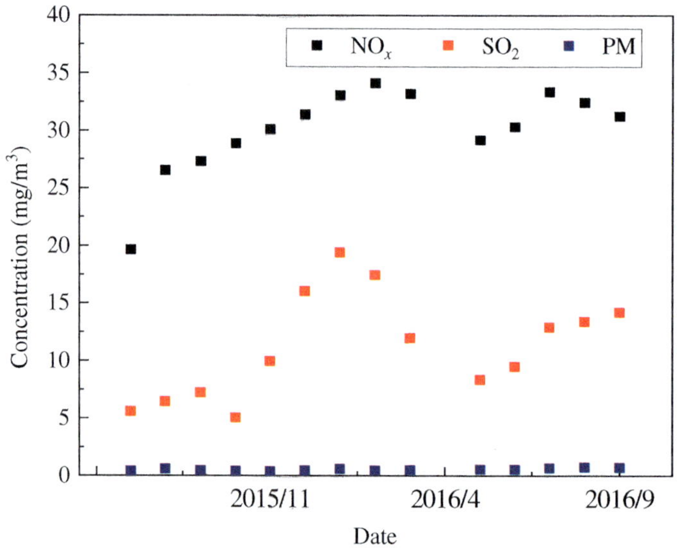

Fig. 4.35 Unit 4 emission status after ultra-low emission upgrade.

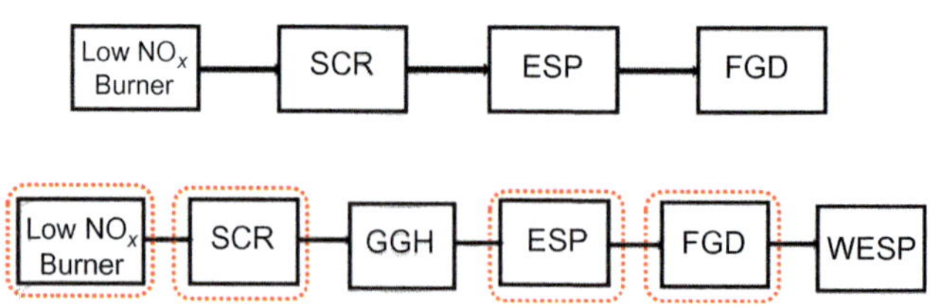

Fig. 4.36 Upgrade roadmap of a power plant in Zhejiang province.

coal-fired type. Limestone gypsum wet flue gas desulfurization technique was used for desulfurization, with a desulfurization efficiency of 95%. The three-chamber four-field ESP also was installed, with a PM collection efficiency of 99.65% [1].

Given the 50% PM removal efficiency of desulfurization system, the comprehensive PM collection efficiency could reach up to 99.92%. With adoption of low-nitrogen burner technology and SCR denitrification (2 layers of catalyst), denitrification efficiency could reach up to 80%. These units were constructed and put into operation in October 2011.

(2) Introduction of upgrade roadmap

In order to further reduce emission concentration of pollutants in flue gas, ultra-low emission technical upgrades were implemented in this power plant in 2013.

PM collection system

Inside the regular ESP, a two-stage tube GGH (gas-gas heat exchanger) was additionally installed. In the meantime, ESP itself was remodeled into a low low-temperature one (with a design flue gas inlet temperature of 86°C), with its power supply replaced with a high-frequency power supply, thus resulting in efficiency increase of more than 99.89%. The WESP was installed after the desulfurization system, with a designed PM removal efficiency of 70%.

Desulfurization system

Its spray system was remolded into one with an interactive spray layer. Besides, an additional layer pallet was installed to form a double-layer pallet. Also, backup circulation pumps were added. Therefore, the desulfurization efficiency was elevated from 95.0% to 98.4%.

Denitrification system

Optimization was performed on the low-nitrogen burners, so that NO_x concentration at the boiler outlet was less than 300 mg/m^3. Besides, a layer of SCR catalyst was added to SCR denitrification system, so that denitrification efficiency was elevated up to 85%. The new catalyst was a newly developed type, working collaboratively with the SCR system to remove Hg and its compounds. The flow chart about this power plant before and after the upgrade is as follows.

(3) Coal quality

The type of coal used by Unit 7 belonged to checked coal, while Unit 8 used the designed coal. During the test period for each unit, coal quality and fuel ratio remained unchanged. Analysis data for coal quality are shown in Table 4.9.

Table 4.9 Analysis data on designed and checked coal

	Unit load	Sulfur (%)	Ash (%)	Volatile matter (%)	Low calorific value (MJ/kg)	Moisture (%)	Fixed carbon (%)	Mercury (10^{-9})
Checked coal	100%	0.75–0.72	16.13–15.75	26.15–26.48	21.87–21.98	14.10–14.10	43.62–43.67	192–278
	75%	0.71–0.68	15.16–15.21	27.64–27.14	22.24–22.68	13.50–12.20	43.70–45.45	
Designed coal	100%	0.45–0.50	11.39–12.06	26.77–26.85	22.37–22.24	15.10–15.40	46.47–46.03	22.1–85.6
	75%	0.31–0.65	9.09–9.72	27.99–27.55	23.06–21.80	14.10–15.70	48.60–48.60	

Adapted from S. Wei, L. Li, Z.M. Cui, J.Y. Wu, H. Mo, Analysis of primary air pollutant emission characteristics and reduction efficiency for ultra-low emission coal-fired power plants based on actual measurement, Electric Power 48 (11) (2015) 131–137.

(4) Analysis on ultra-low emission operational characteristics

After optimization of both low-nitrogen burners and SCR denitrification equipment, the NO_x emission concentration and denitrification efficiency status in this plant denitrification is hereby shown in Fig. 4.37. During the test period, the flue gas temperature at four flue gas passage inlets of the two units ranged between 310°C and 320°C (average: 314.88°C), the moisture content ranged between 6.2% and 6.8% (with an average value of 6.41%). Fig. 4.37 indicates that, after optimization of the low-nitrogen burners and under various working loads, NO_x concentration at Unit 7 boiler outlet fell within 45–288 mg/m^3, while NO_x concentration at Unit 8 fell within 225–283 mg/m^3. After SCR denitrification, NO_x emission concentrations at two units stack inlets fell within the range of 18.43–31.6 mg/m^3 and 18.56–34.0 mg/m^3, respectively.

SO_2 emission concentration and desulfurization efficiency are shown in Fig. 4.38. As to Unit 7 when operating under various loads, it could be seen from the Figure that, after the desulfurization system ultra-low emission upgrade, SO_2 concentration at the desulfurization absorption system inlet ranged between 1451 and 1614 mg/m^3, and at the outlet ranged between 9.54 and 1.01 mg/m^3, with a desulfurization efficiency of 98.95%–99.34%. As to Unit 8 operating under various loads, SO_2 concentration at the desulfurization absorption system's inlet ranged between 830 and 978 mg/m^3, and at the outlet ranged between 7.69

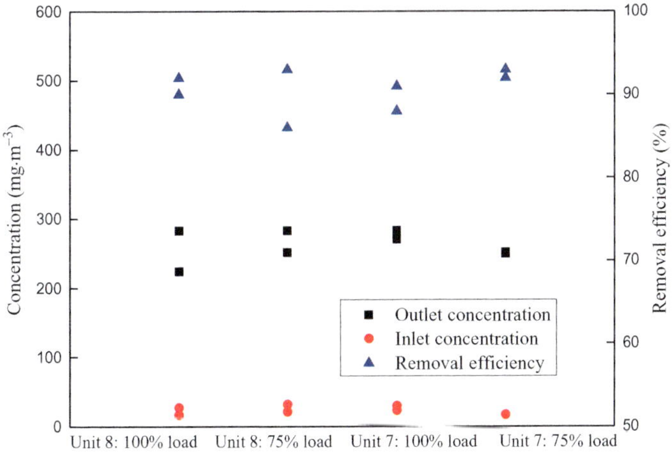

Fig. 4.37 NO_x concentration and removal efficiency after upgrade. *(Redrawn from S. Wei, L. Li, Z.M. Cui, J.Y. Wu, H. Mo, Analysis of primary air pollutant emission characteristics and reduction efficiency for ultra-low emission coal-fired power plants based on actual measurement, Electric Power 48 (11) (2015) 131–137.)*

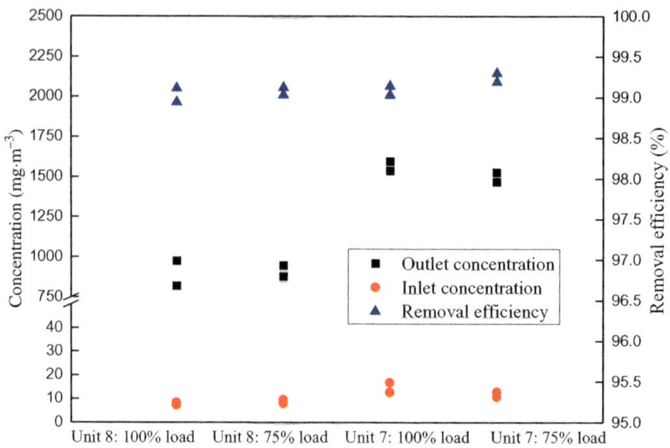

Fig. 4.38 SO_2 concentration and removal efficiency after upgrade. *(Redrawn from S. Wei, L. Li, Z.M. Cui, J.Y. Wu, H. Mo, Analysis of primary air pollutant emission characteristics and reduction efficiency for ultra-low emission coal-fired power plants based on actual measurement, Electric Power 48 (11) (2015) 131–137.)*

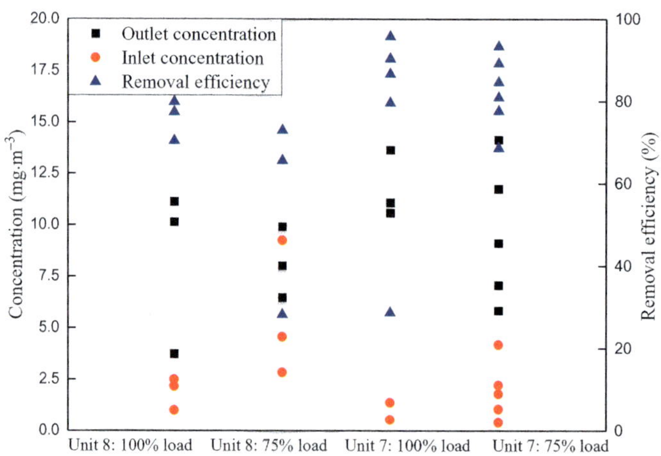

Fig. 4.39 PM concentration and removal efficiency after upgrade. *(Redrawn from S. Wei, L. Li, Z.M. Cui, J.Y. Wu, H. Mo, Analysis of primary air pollutant emission characteristics and reduction efficiency for ultra-low emission coal-fired power plants based on actual measurement, Electric Power 48 (11) (2015) 131–137.)*

and 9.51 mg/m^3, with a desulfurization efficiency of 98.88%–99.18%. The emission concentration of the ultra-low emission system therefore met the designed limit requirements.

PM emission concentration and removal efficiency are hereby shown in Fig. 4.39. When Unit 7 was operating under various loads, PM concentration at desulfurization absorption system's inlet ranged between 5.95 and

13.99 mg/m^3, while PM concentration at both outlets of the absorption system and the WESP ranged between 1.10 and 9.30 mg/m^3, with a PM collection efficiency of 69.27%–95.27%. As to Unit 8 operating under various loads, PM concentration at desulfurization absorption system inlet ranged between 3.70 and 11.16 mg/m^3, while PM concentration at both outlets of the absorption system and the WESP ranged between 1.10 and 9.30 mg/m^3, with a PM collection efficiency of 28.94%–79.39%. PM test results indicated that requirements of ultra-low emission limit of 5 mg/m^3 as designed were well met.

4.3 Emissions control technology based on hybrid ESP-FF

4.3.1 Jiaozuo Power Plant

(1) Introduction of power units before upgrade

660-MW ultra-supercritical boiler of Jiaozuo Power Plant belonged to Model SG-2024/26.15-M6009, manufactured by Shanghai Boiler Works Co., Ltd. It was a single corrugated furnace DC π-type boiler, with variable pressure operationing ability, four-corner tangential firing, and one intermediate reheat function.

PM collection system

Hybrid ESP + FF were adopted for flue gas PM collectors, with a designed removal efficiency of $\geq$99.95%, and a design value of $\leq$10 mg/m^3 for flue gas emission at outlets of the PM collector. In this case, average PM emission could be continuously maintained below 10 mg/m^3.

Desulfurization system

Limestone gypsum desulfurization technique was adopted for SO_2 removal, designed with a one-furnace, one-absorption system, and a five-layer spray, but without bypass or GGH system. Designed desulfurization efficiency was more than 98% under 100% flue gas conditions. The average emission concentration could be maintained at a level below 35 mg/m^3. Inside the desulfurization system, some subsystems were also installed, such as pulping, gypsum dehydration, and desulfurization treatment of wastewater.

Denitrification system

Besides employment of low nitrogen burners, SCR (selective catalytic reduction) technology was also adopted for flue gas denitrification. Each boiler was equipped with two sets of a denitrification reactor and a honeycomb catalyst, accompanied with 2 + 1 configuration of reactors and use of liquid ammonia as a reducing agent. The design denitrification efficiency was more than 80%, with average emission concentration being stably maintained below 50 mg/Nm^3.

(2) Introduction to upgrade roadmap
Hybrid ESP-FF was adopted for the initial flue gas PM collectors. Three independent electric fields were introduced for ESP electric fields, all equipped with high-frequency power supplies, with high reliability and stability. PM collection efficiency in electric fields could reach up to more than 85%. It not only possessed energy-saving ability, but also high PM collection efficiency. Furthermore, two areas were specifically designated for fabric filters at the rear of the PM removal system. PM concentration at the stack outlet was around 10–15 mg/m^3. The Hybrid ESP-FF was in good operation ever since it was put into operation. All the components running in the electric field section and fabric filter section were in a stable condition. As a result, the upgrade to ESP + FF was not necessary.

Despite the adoption of wet limestone gypsum flue gas desulfurization equipment, there were not only PM but also a large amount of desulfurization products, including limestone in the emitted flue gas. This proved that it was not realistic to expect to reduce the PM emission concentration in the flue gas merely by controlling the PM concentration at the inlet of the desulfurization equipment. Furthermore, the secondary pollution of the desulfurization equipment also contributed to the excessive pollutant concentration in the emitted flue gas. Thus, the conclusion was that the upgrade for PM only needs to improve the desulfurization absorption system to increase PM removal efficiency, in order to ensure that PM emission concentration at its outlet be kept within a certain level. By this way, the PM emission concentration at the FGD outlet could meet the ultra-low emission requirement of 5 mg/m^3.

The ultimate solution adopted by this power plant was to install an integrated ultra-low emission desulfurization particle trap (a high-efficiency demister of FGD), which utilized a three-layer roof ridge-type high-efficiency demister. And a uniform airflow distribution adjuster was additionally installed at the demister inlet.

The flow field in the absorption system was able to be adjusted through installation of airflow uniform distribution adjusters at the front portion of the demister. The amount of slurry contained in PM was able to be lowered when it entered the high-efficiency demister. In the running operation, when the demisting effect of the high-efficiency demister increased, the amount of liquid water and gypsum contained in flue gas at the absorption system outlet reduced. Furthermore, this adjusted the flue gas flow velocity inside the blades of the high-efficiency demister, and the performance of the high-efficiency demister was improved.

For the absorption system, single-absorption system and double-zone technology were adopted. Inside the slurry tank, slurry and liquid were separated. Fresh limestone slurry was injected directly into the circulation pump inlet zone. The internal slurry in absorption system was divided into an oxidation crystallization zone and an absorption transport zone. Furthermore, the upgrade was conducted on the demister and the oxidizing air network, in order to enhance the sulfur dioxide absorption effect.

The coal used came from Shenhua bituminous coal, having the quality of low sulfur, low ash content, and high volatile content. Ever since the start of operation, with adoption of a newly developed low nitrogen burner technology, the NO_x concentration at the boiler economizer outlet was able to stay below 200 mg/m^3. At the denitrification system outlet, it was able to stay constantly below 50 mg/m^3. In order to enhance denitrification, the amount of denitrification ammonia was increased to a great extent during the operation. The flow chart about Jiaozuo Power Plant before and after the upgrade is as follows (Fig. 4.40).

(3) Coal quality

The coal used by the two coal-fired power units was from Boertai Coal, Bulianta Coal, and Yima Coal. Tests on the quality of coal as admitted were conducted in June, July, August, September, October, November, and December of 2016. The specific coal quality is listed in Table 4.10 (Fig. 4.41).

From Table 4.10, it can be seen that moisture and volatile matter were relatively stable, while ash content and sulfur content varied widely. The minimum ash content value during the statistical period was 7.7%, the maximum value was 28.44%. The minimum sulfur content value was 0.37%, and the maximum value was 1.23%. According to Table 4.8, the coal quality analysis from June to December 2016 showed that the average value of total moisture on as-received basis was 11.77%, the average value of ash content on as-received basis was 19.51%, and the average value of sulfur content on as-received basis was 0.64%. The ash changes in different months were also relatively large.

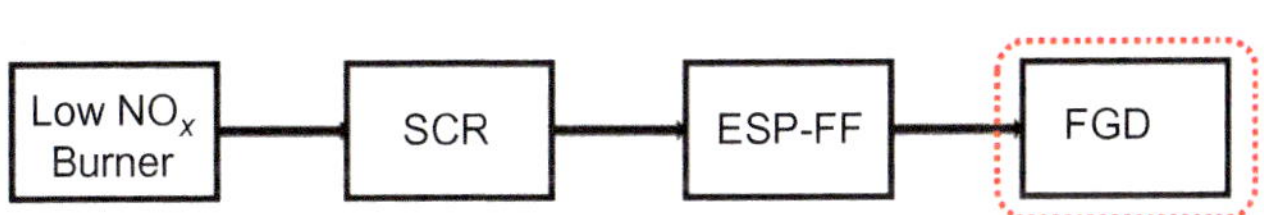

Fig. 4.40 Jiaozuo Power Plant upgrade roadmap.

Table 4.10 Analysis data on coal quality for Jiaozuo Power Plant

Time	Coal type	Monthly statistics	Moisture (%)	Ash (%)	Sulfur (%)	Volatile matter (%)	Low calorific value (MJ/kg)
2016.06	Lignite	Maximum	14.5	15.34	0.57	38.84	24.99
		Mean	13.0	11.15	0.47	38.12	23.79
		Minimum	12.0	7.70	0.37	37.04	22.30
2016.07	Lignite	Maximum	12.9	19.93	0.64	39.19	24.87
		Mean	12.1	16.91	0.56	38.13	22.04
		Minimum	10.0	11.79	0.39	37.13	20.91
2016.08	Lignite	Maximum	12.9	22.80	0.80	39.27	23.25
		Mean	11.6	18.82	0.64	37.99	21.47
		Minimum	10.6	13.72	0.44	36.79	20.07
2016.09	Lignite	Maximum	11.9	23.52	0.91	38.50	22.53
		Mean	10.7	19.81	0.69	37.23	21.47
		Minimum	10.1	16.68	0.57	35.24	20.05
2016.10	Lignite	Maximum	13.1	28.44	0.98	40.22	21.80
		Mean	11.4	23.67	0.68	37.09	19.53
		Minimum	10.0	19.59	0.27	34.34	17.50
		Maximum	13.7	28.44	1.23	39.80	21.27
2016.11	Lignite	Mean	11.8	23.67	0.76	37.16	19.76
		Minimum	10.9	19.59	0.40	34.31	17.79
		Maximum	12.6	25.87	0.84	38.09	22.92
2016.12	Lignite	Mean	11.8	21.31	0.68	36.24	20.67
		Minimum	11.2	14.06	0.53	35.15	19.32

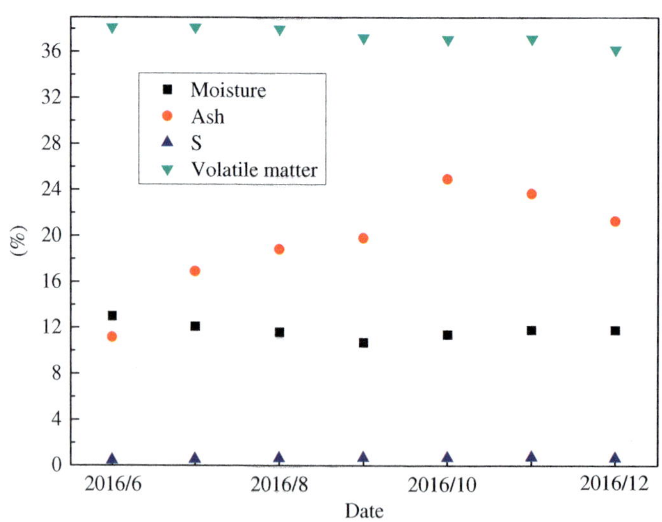

Fig. 4.41 Coal quality analysis in Jiaozuo power plant.

(4) Operation analysis of ultra-low emission

Figs. 4.42 to 4.45 show pollutant emission data at different times before and after the ultra-low emission upgrade of Jiaozuo Power Plant. For Unit 1, before the ultra-low emission upgrade, the PM and SO_2 emission concentration exceeded 10 and 35 mg/m^3. The concentration of

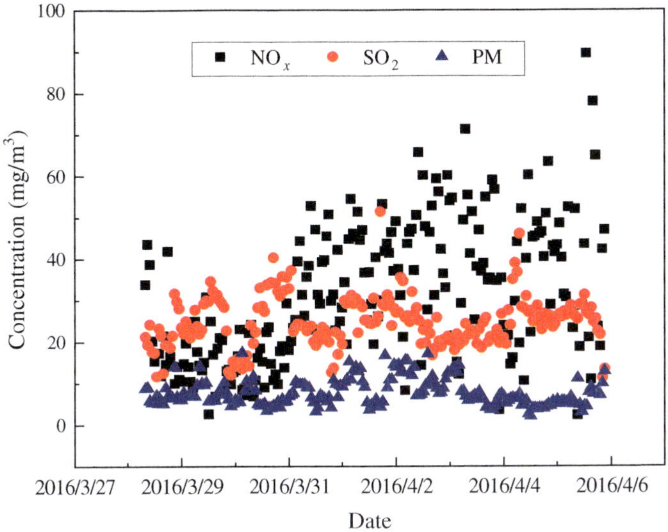

Fig. 4.42 Emission before ultra-low emission upgrade of Unit 1.

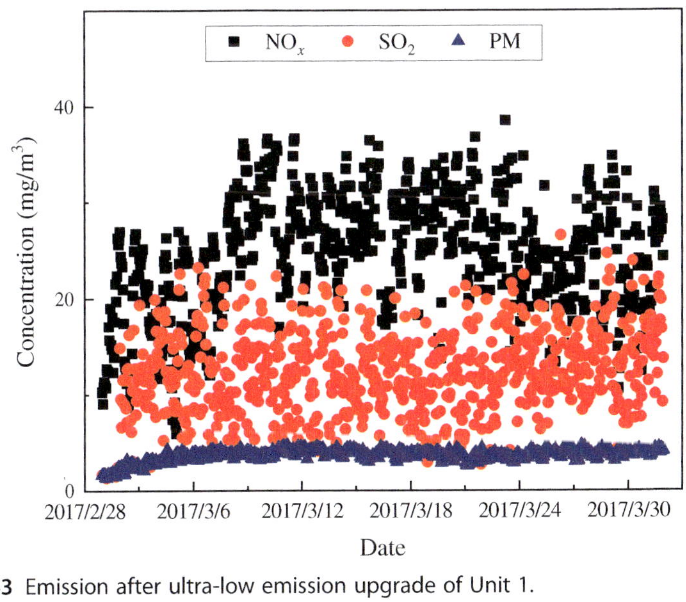

Fig. 4.43 Emission after ultra-low emission upgrade of Unit 1.

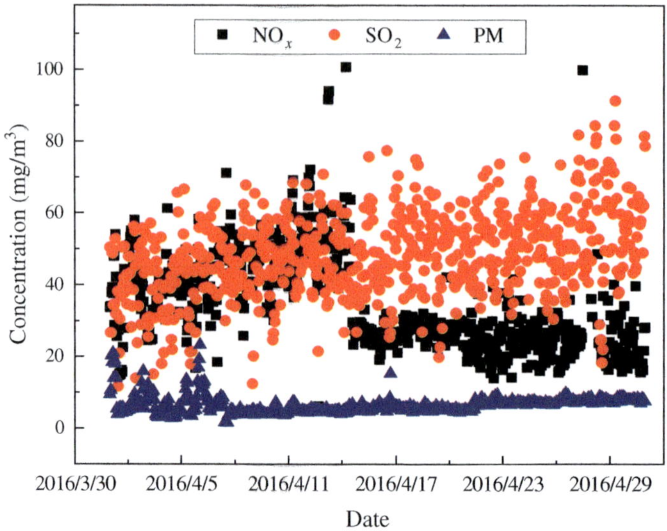

Fig. 4.44 Emission before ultra-low emission upgrade of Unit 2.

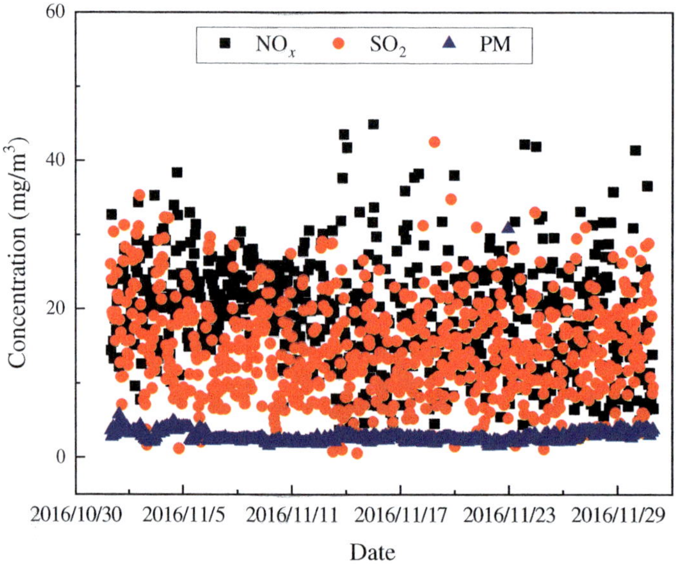

Fig. 4.45 Emission after ultra-low emission upgrade of Unit 2.

NO_x emission fluctuated obviously, and it exceeded 50 mg/m^3 for a considerable amount of time. Within half a year after the upgrade, the average emission concentration of PM was 2.85 mg/m^3, the average emission concentration of SO_2 was 13.80 mg/m^3, and the average emission concentration of NO_x was 18.73 mg/m^3. The average emission concentration of

PM, SO_2, and NO_x after the upgrade of coal-fired units in Jiaozuo Power Plant was lower than the limit of atmospheric pollutant emission concentration of gas-fired units, which met the ultra-low emission standards. The effect of the upgrade emission reduction was remarkable.

NO_x, SO_2, and PM emission concentration were relatively high before the ultra-low emission upgrade of Unit 2. The PM emission concentration exceeded 10 mg/m^3, SO_2 exceeded 35 mg/m^3, and its concentration was often between 60 and 80 mg/m^3. In contrast, only in few cases, the NO_x emission concentration was above 50 mg/m^3. The emission of NO_x, SO_2, and PM after the ultra-low emission upgrade was significantly reduced, especially the SO_2 emission concentration decreased sharply, and the average concentrations were 19.39, 14.62, and 2.84 mg/m^3, respectively, which met the emission standards.

4.3.2 Qinhuangdao Power Plant

(1) Introduction of the unit before upgrade

Qinhuangdao Power Generation Co., Ltd. had two 215-MW and two 320-MW condensing generator sets with a total capacity of 1070 MW.

PM collection system

The PM collection mode of the four units was hybrid ESP + FF.

Desulfurization system

Seawater desulfurization was adopted for all the gas SO_2 removal systems of the four units, including: (1) absorption system, (2) seawater supply and drainage system, (3) water media pipe heat exchanger (fluoroplastic HDS technology). Units 1 and 2 were two boilers and one FGD method, and Units 3 and 4 were one boiler with one FGD method.

Denitrification system

The denitrification methods were equipped with low nitrogen burners and selective catalytic reduction, of which SCR adopted 2 + 1 mode. The ammonia supply was provided for 4 units through the main pipe from the ammonia storage area.

(2) Introduction to upgrade roadmap

Flow distribution devices and particle precollection devices were installed at the inlet of the ESP + FF, and the PM emission concentration at the outlet of the PM collector was controlled below 15 mg/m^3. A first-stage high-efficiency PM collector was installed in the desulfurization absorption system, and a second-stage high-efficiency demister was installed in the flue of the desulfurization absorption system. The PM emission concentration at stack inlet was controlled below 5 mg/m^3. The seawater distributor (spray layer main pipe, branch pipe and nozzle, etc.) was modified accordingly. Rotary GGH was a replacement by water media pipe heat exchanger and flat plate-type high-efficiency demister. Furthermore, damaged and deformed filler was replaced in time, and its height was increased.

The upgrade solution for NO_x control was adding an elevated catalyst to the SCR reserved layer, which achieved an ultra-low emission NO_x emission concentration below 50 mg/m^3.

Therefore, the overall upgrade roadmap was: improvement denitrification efficiency upgrade + removal PM by flow distribution and particle precollection + improvement desulfurization + high-efficiency demister upgrade.

The flowchart before and after the upgrade of Qinhuangdao Power Plant is as follows (Fig. 4.46).

(3) Coal quality

The designed coal was Shanxi Jinbei bituminous coal, and the coal quality from January 2014 to September 2016 was tested. The specific coal quality is shown in Table 4.11. In addition, the coal quality from January 2014 to September 2016 is plotted and analyzed on a monthly basis (Fig. 4.47).

The average value of total moisture content on the as-received basis was 10.44%, with a limited moving range. And the average value of the ash content on as-received basis was 21.89%, while the change fluctuated significantly. The average value of sulfur content on as-received basis was 0.58%, the average value of the volatile matter content was 26.54%, both of them generally remained at a constant level without much fluctuation.

(4) Operation analysis of ultra-low emission

Figs. 4.48 and 4.49 show the characteristics of PM, SO_2, and NO_x emissions from flue gas at different times before and after the ultra-low

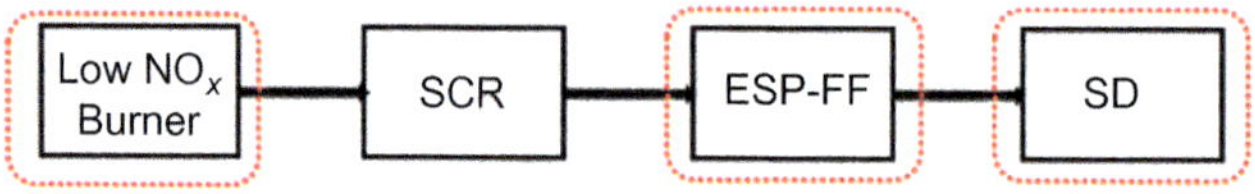

Fig. 4.46 Upgrade roadmap of Qinhuangdao Power Plant.

Table 4.11 Coal quality analysis in Qinhuangdao Power Plant

Time	Coal	Moisture (%)	Ash (%)	Sulfur (%)	Volatile matter (%)	Net calorific value (MJ/Kg)
2014	Bituminous	8.0	26.38	0.77	25.94	19.82
2015	Bituminous	11.27	20.34	0.52	26.88	20.56
2016	Bituminous	12.06	18.95	0.46	26.78	20.77
Average	Bituminous	10.44	21.89	0.58	26.54	20.38

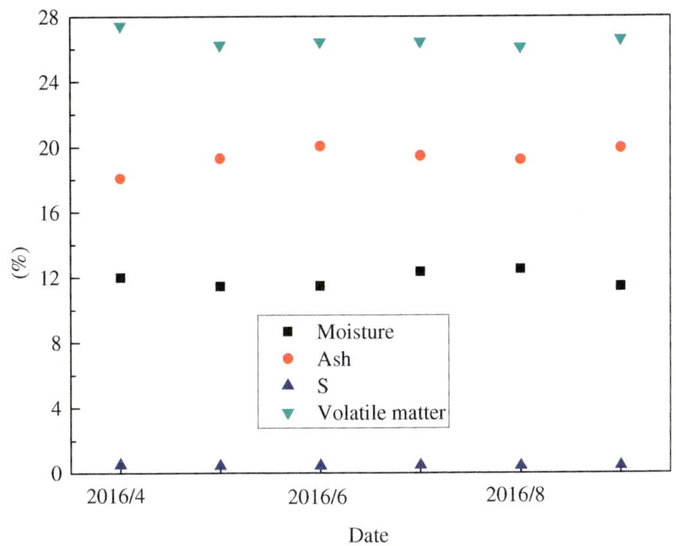

Fig. 4.47 Coal quality analysis in Qinhuangdao Power Plant.

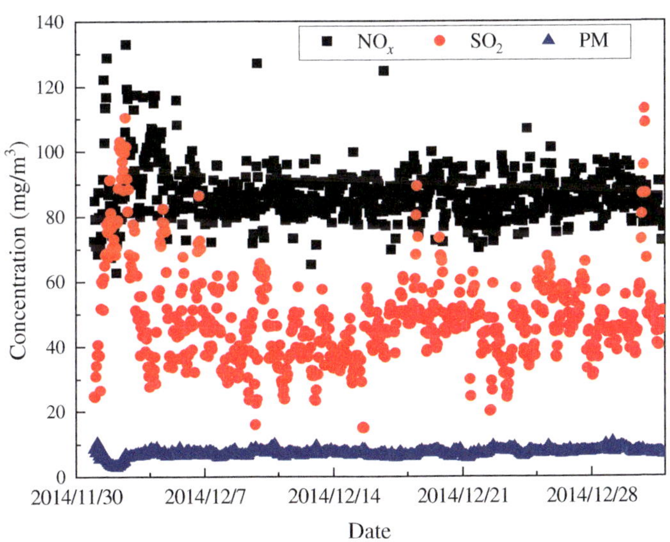

Fig. 4.48 Emission before ultra-low emission upgrade of Unit 1.

emission of Unit 1 of Qinhuangdao Power Plant. Before the ultra-low emission upgrade of Unit 1, the NO_x emission concentration was between 80 and 100 mg/m^3 in most cases; the concentration of SO_2 was mostly between 30 and 60 mg/m^3. PM concentration ranged from 5 to 10 mg/m^3. After ultra-low emission upgrade, NO_x emission concentration of Unit 1 was in the range of 30∼40 mg/m^3 in most cases; the

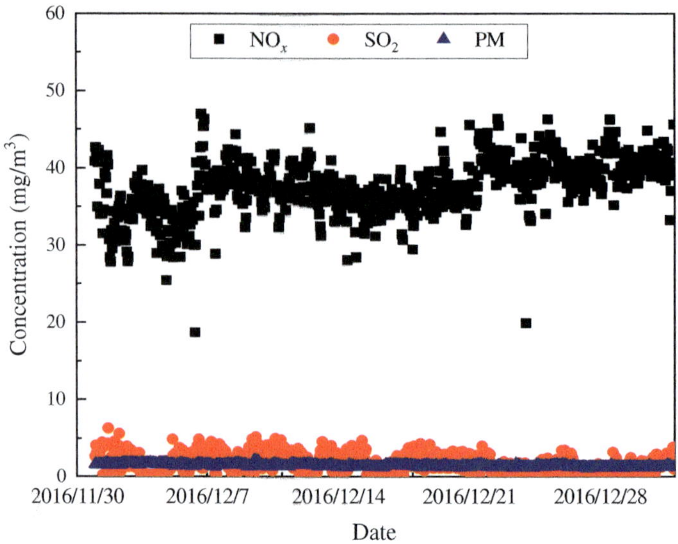

Fig. 4.49 Emission after ultra-low emission upgrade of Unit 1.

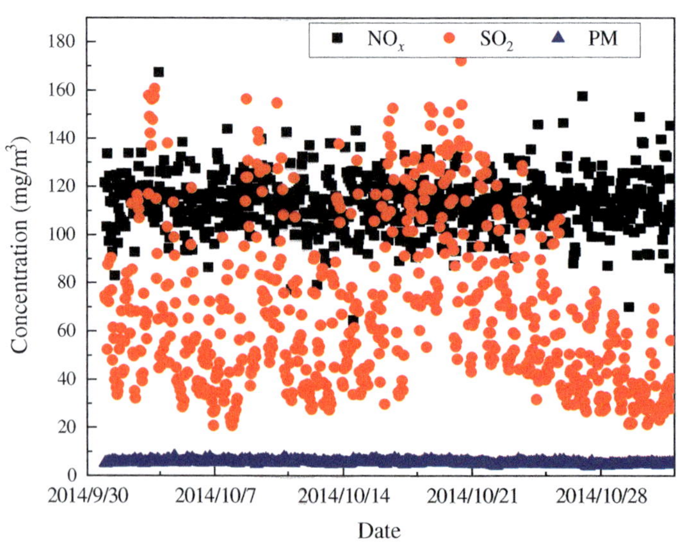

Fig. 4.50 Emission before ultra-low emission upgrade of Unit 3.

concentration of SO_2 decreased obviously, most of which were 0~5 mg/m^3, and the maximum value was only 6.25 mg/m^3; the PM concentration was less than 5 mg/m^3, and the peak value was 2.12 mg/m^3.

Figs. 4.50 and 4.51 show the characteristics of PM, SO_2, and NO_x emissions from flue gas at different times before and after the ultra-low

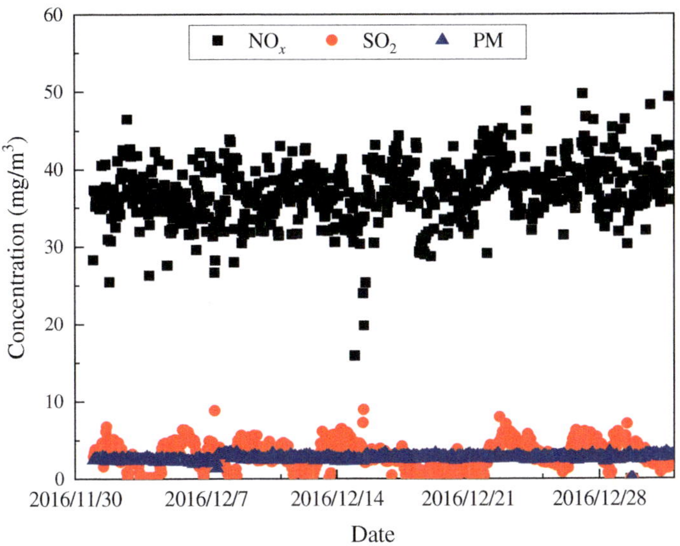

Fig. 4.51 Emission after ultra-low emission upgrade of Unit 3.

emission upgrade of Unit 3 of Qinhuangdao Power Plant. Before the ultra-low emission upgrade of Unit 3, the NO_x emission concentration was between 100 and 130 mg/m^3 in most cases; the concentration distribution of SO_2 was relatively dispersed, some of which met the ultra-low emission standard, but some of which were much higher than the limited value that emission standard specified. The PM emission concentration fluctuated at 5 mg/m^3, plus or minus 1 mg/m^3. After the ultra-low emission upgrade, NO_x emission concentration of Unit 3 was in the range of 30∼45 mg/m^3 in most cases; the emission concentration of SO_2 was relatively stable in the range of 0∼10 mg/m^3; the PM concentration was less than 5 mg/m^3.

Figs. 4.52 and 4.53 show the PM, SO_2, and NO_x emissions characteristics of the flue gas at different times before and after the ultra-low emission upgrade of Unit 4 of the power plant. Before the ultra-low emission upgrade of Unit 4, the NO_x emission concentration was in the range of 100∼140 mg/m^3 in most cases, the concentration distribution of SO_2 was relatively dispersed, which fluctuated significantly from 40 to 160 mg/m^3; the PM emission concentration fluctuated within the range of 10 mg/m^3, plus or minus 2 mg/m^3. After the ultra-low emission upgrade, the NO_x emission concentration of Unit 4 was mostly in the range of 30–40 mg/m^3; the SO_2 emission concentration was in the range of 0–10 mg/m^3; the PM concentration was less than 5 mg/m^3.

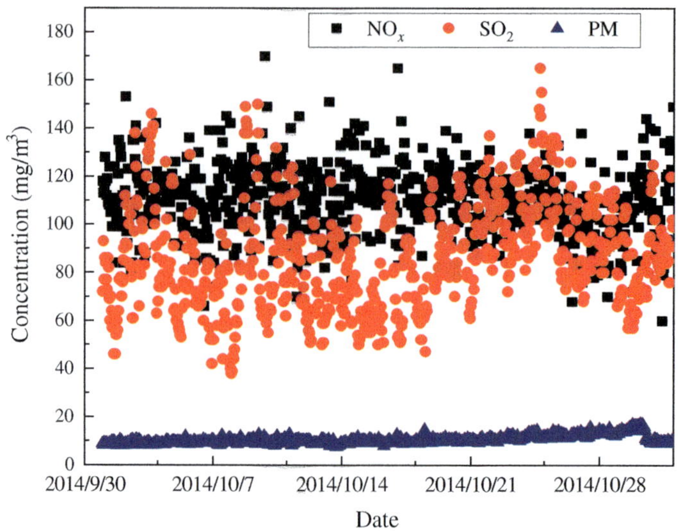

Fig. 4.52 Emission before ultra-low emission upgrade of Unit 4.

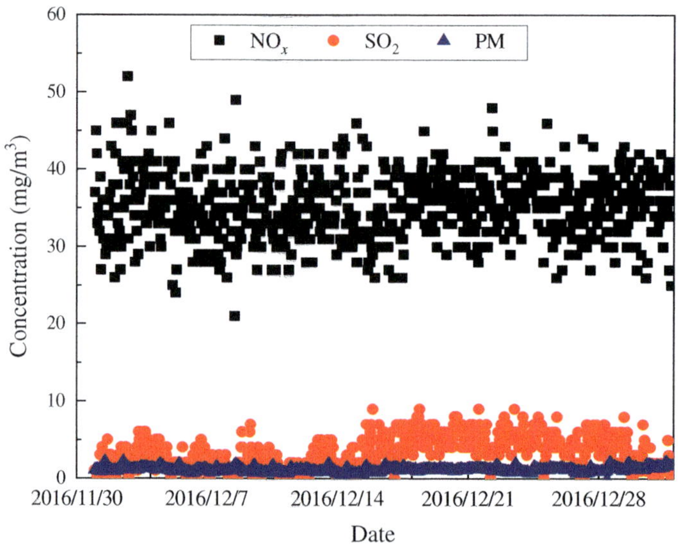

Fig. 4.53 Emission after ultra-low emission upgrade of Unit 4.

By comparing the pollutant discharge concentrations before and after the ultra-low emission upgrade of the same unit, it could be seen that the ultra-low emission upgrade had a more obvious effect on the removal of NO_x and SO_2. It could be seen from the upgrade plan that the addition of the elevated catalyst in the SCR reserved layer could effectively reduce the NO_x emission concentration and achieve the ultra-low emission standard. The upgrade of the desulfurization and efficiency improvement was also very successful. The rotary GGH was replaced by a water media pipe heat exchanger and a high-efficiency demister was added to effectively reduce the SO_2 concentration. It could not be maintained in a more stable state, and the fluctuation range was not obvious. In addition, through these figures, it could be seen that after the ultra-low emission upgrade, the PM, SO_2, and NO_x emission concentrations during the sampling time were basically lower than 5, 30, 40 mg/m^3, which was in line with the ultra-low emission10, 35, 50 mg/m^3 standard requirement.

4.3.3 Garden Power Plant

(1) Introduction of the unit before upgrade

The Garden Power Plant was equipped with a 4 × 660 MW single-shaft condensing steam turbine unit, which was equipped with a 4 × 2236 t/h supercritical and intermediate reheat coal-fired boiler. It was constructed with two-sleeve double steel inner tube stacks with a height of 210 m and a single inner tube diameter of 7.6 m, and one stack was used for each two boilers to discharge flue gas.

PM collection system

The PM collection system used an ESP with four chambers and a five-electric field. High-frequency power supply was adopted in all the five electric fields, and the PM precharged microparticle collection enhancement technology was also provided. It was designed to ensure the PM collection efficiency of 99.92%, plus with desulfurization system PM collection efficiency of 50%, the total PM collection efficiency designed of 99.93%.

Desulfurization system

The desulfurization system adopted the wet limestone gypsum process, including absorbent preparation system (including limestone supply, slurry preparation, and supply system), flue gas system, absorption system, gypsum dehydration system, emergency discharge system, process water system, desulfurization wastewater treatment system. The spray layer in the desulfurization absorption system adopted 4 + 1 mode, and there was no GGH and desulfurization flue gas bypass. The design desulfurization efficiency was 95%.

Denitrification system

SCR process was adopted in the flue gas denitrification system. One boiler was equipped with one reactor, and the denitrification flue gas system was not bypassed. The denitrification efficiency was not less than 80% according to the designed coal and the checked coal. The catalyst was arranged in 2+1 layers mode, one layer was reserved. The designed NO_x concentration of flue gas emission was less than 100 mg/m^3, and the denitrification reducing agent was liquid ammonia.

(2) Introduction to upgrade roadmap

The ESP with double chambers and a five-electric field of the PM collection system could ensure that the PM collection efficiency was maintained above 99.92%. The PM concentration of the existing ESP+FF was lower than 20 mg/m^3, so the PM emission concentration of the stack inlet could be directly reduced by 5 mg/m^3 through the desulfurization system. Therefore, the ESP was not included in the unit upgrade plan.

The desulfurization device adopted a wet limestone gypsum scheme of one boiler and one absorber mode. The absorption system adopted a five-layer spray, two-stage demister spray design, and the slurry pipeline adopted a double circuit to ensure the desulfurization efficiency was greater than 95%. However, the spray effect of the original absorption system could not meet the new desulfurization efficiency requirements. It was necessary to upgrade the details of the gas flow distribution in the absorber of the existing absorption system, the contact time and contact area of the gas-liquid and the spray slurry to achieve 98.25 % of the design desulfurization efficiency. After the replacement of the high-efficiency demister, the three-story roof ridge structure was selected, and the demister modules were differentiated to adapt to the flow field in the absorption system.

In the denitrification system, the SCR catalyst of the Garden Power Plant was designed according to "2+1" mode, and a total of three layers of catalysts were arranged. A layer of catalysts and an acoustic soot blower were added, and the denitrification reducing agent was liquid ammonia. The backup catalyst was installed during the maintenance period of the unit, and the layer was added with a sonic soot blower at this time. The denitrification flow field was optimally adjusted, and the guide plate was increased, which met the new requirement of controlling NO_x emission less than 50 mg/m^3. The flow chart before and after the upgrade of the Garden Power Plant is as follows (Fig. 4.54).

(3) Coal quality

Coal-fired coal was used in the lignite coal of No. 1 Mine and the open pit mine of Dananhu No. 2 Mine in Dananhu Mining Area. The coal quality of the boiler in 2016 was tested, and the specific coal quality is shown in Table 4.12 (Fig. 4.55).

Fig. 4.54 Upgrade roadmap of Huanyuan Power Plant.

Table 4.12 Coal quality analysis in Huayuan Power Plant

Month	Coal	Moisture (%)	Ash (%)	Sulfur (%)	Volatile matter (%)	Net calorific value (MJ/Kg)
1	Lignite	25.34	23.64	0.33	23.52	13.43
2	Lignite	25.02	24.54	0.28	24.05	13.15
3	Lignite	25.06	24.16	0.30	23.42	13.27
4	Lignite	26.53	20.54	0.29	23.79	13.69
5	Lignite	26.35	19.42	0.29	24.43	14.22
6	Lignite	26.40	19.91	0.3	23.88	13.97
7	Lignite	27.14	18.91	0.26	29.34	13.66
8	Lignite	26.79	20.55	0.27	23.63	13.70
9	Lignite	26.72	20.64	0.28	23.86	13.62
10	Lignite	25.96	22.01	0.29	24.07	13.85
11	Lignite	26.90	19.03	0.31	24.49	14.22
12	Lignite	26.88	22.16	0.27	23.01	13.51
Average	Lignite	26.26	21.29	0.29	24.29	13.69

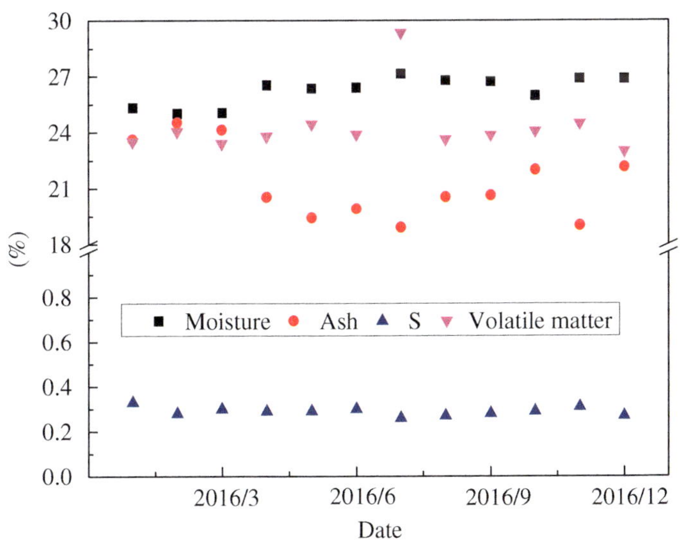

Fig. 4.55 Coal quality analysis in Huayuan Power Plant.

At the same time, in order to observe the change of coal quality more intuitively, the data above were plotted and analyzed. The average value of total moisture content on as-received basis was 26.26%, and the total moisture content on as-received basis was less than 1% except in February and March. The ash content on as-received basis from coal burning was relatively high in the measurement of the first three months, reaching a peak of 24.5%. However, that generally maintained at about 20% in later period. The sulfur content on as-received basis and the volatile matter content on as-received basis remained basically at 0.3% and 23.5%, except the volatile matter content from coal burning in July reached 29.34%. The coal quality used in the coal-fired units of the Huayuan Power Plant was within the acceptable range of operating time before and after the ultra-low emission upgrade, and there were little changes in the coal quality before and after the upgrade were little.

(4) Operation analysis of ultra-low emission

Figs. 4.56 and 4.57 show the PM, SO_2, and NO_x emissions characteristics of the flue gas monitored hourly before and after the ultra-low emission upgrade of the Unit 3 in Huayuan Power Plant. Before the ultra-low emission upgrade, the NO_x concentration of Unit 3 of Huayuan Power Plant basically maintained at 15–40 mg/m^3, the SO_2 concentration varied between 30 and 60 mg/m^3, and the PM concentration was also 10–15 mg/m^3. NO_x met ultra-low emission standards, while SO_2

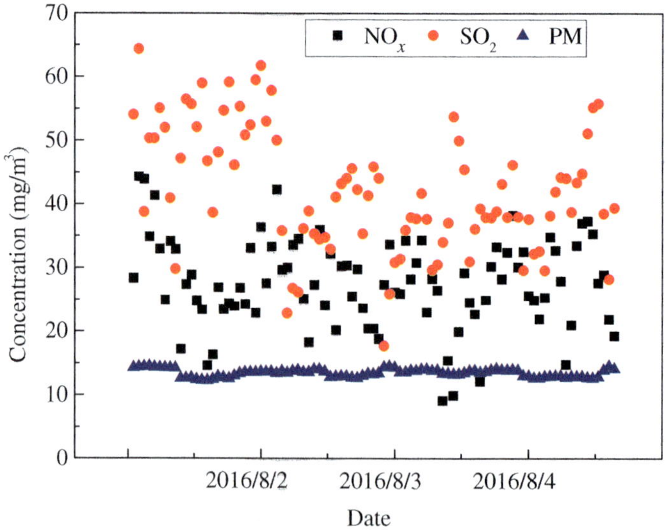

Fig. 4.56 Emission before ultra-low emission upgrade of Unit 3.

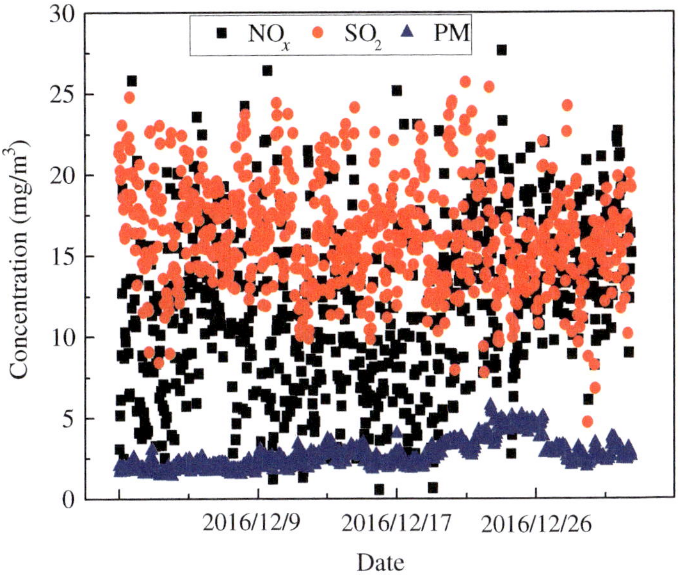

Fig. 4.57 Emission after ultra-low emission upgrade of Unit 3.

and PM did not reach the ultra-low emission standards. In contrast, after the ultra-low emission upgrade, the pollutant emission concentration of Unit 3 had dropped significantly, the NO_x concentration was less than 25 mg/m^3, the SO_2 concentration range was 10–25 mg/m^3, and the PM concentration was reduced to 5 mg/m^3 as below. Compared with the emission before and after the ultra-low emission upgrade, it shown that the ultra-low emission upgrade had more obvious effects on SO_2 and PM removal. Little effect on NO_x and may attribute to the denitrification system before the unit upgrade was equipped with a good performance low nitrogen burner technology and SCR equipment. After the upgrade, the synergistic desulfurization absorption system minimized the concentration of SO_2 in the flue gas. The high-efficiency demister played a significant role in removing SO_2, and had a synergistic effect on PM. It showed from the figure that after the ultra-low emission upgrade, the concentration of PM, SO_2, and NO_x emissions during the monitoring period was lower than 5, 25, 25 mg/m^3, which was much lower than the ultra-low emission of standard of 10, 35, 50 mg/m^3.

Fig. 4.13 is the acceptance test data of the Unit 3 in 2016 for Garden Power Plant after the completion of the upgrade project. It showed that the average emission concentration of PM, SO_2, and NO_x was lower than the limit of atmospheric pollutant emission concentration of the gas

Table 4.13 Operation data after ultra-low emission upgrade of Unit 3 of Huayuan Power Plant

	PM (mg/m^3)			NO_x (mg/m^3)			SO_2 (mg/m^3)		
Month	Month average	Daily min	Daily max	Month average	Daily min	Daily max	Month average	Daily min	Daily max
11	2.4	1.5	5.8	15.3	9.9	25.5	16.1	11.3	23.7
12	2.7	1.8	4.6	11.5	6.6	16.3	16.1	12.4	23.1

turbine unit, which was in line with the ultra-low emission standard, and the effect of upgrading emission reduction was remarkable (Table 4.13).

From the ultra-low emission practice of Jiaozuo Power Plant, Qinhuangdao Power Plant, and Garden Power Plant, it is found that the denitrification performance was enhanced on the original basis in most cases. In the PM removal upgrade, there is little change on the ESP-FF system. For the secondary carrying in the desulfurization absorption system, the demister was strengthened with a simple upgrade on existing FGD. In the end, ultra-low emission was achieved. For this roadmap, it is easy to implement with less construction work.

4.4 Emissions control technology based on Dry FGD (DFGD)

4.4.1 Hequ CFB Power Plant

(1) Introduction to the implementation of ultra-low emission

Two new 350-MW supercritical CFB (circulating fluidized bed) units were constructed at Hequ CFB Power Plant. The boiler was a type DG1150/25.4-II1 supercritical once-through circulating fluidized bed manufactured by Dongfang Boiler Co., Ltd.

The construction of ultra-low emission was simultaneously implemented with the CFB unit. An ESP with double chambers and first electric field pre-electrostatic precipitator, which designed PM collection efficiency greater than 80%, was equipped before the desulfurization system, and an FF efficiency of not less than 99.9% was adopted after the desulfurization system.

The SO_2 removal system adopted a combining dry spray calcium and semidry process of flue gas desulphurization. The desulfurization system adopted one boiler with one absorber arrangement. The desulfurization efficiency inside the furnace was not less than 90%, and the desulfurization efficiency outside the furnace was not less than 76%. The designed comprehensive SO_2 removal efficiency was not to be lower than 97.6%.

The ultra-low emission removal roadmap of the circulating fluidized bed unit was quite different from that of the pulverized coal boiler. CFB power plant often adopts selected selective noncatalytic reduction (SNCR) technology for NO_x control. CFB boilers had lower furnace temperatures than pulverized coal-fired boilers; thus, there was very little thermal NO_x production, at the same time, the furnace temperature was in accordance with the temperature window in which the reducing agent reacted with NO_x. The unit fully utilized the low bed temperature and low bed pressure combustion technology to control the amount of NO_x produced. Moreover, the operating temperature of the circulating fluidized bed is suitable for the desulfurization technology of the fuel combustion process. Therefore, the circulating fluidized bed unit adopted added limestone in the furnace for desulfurization during combustion. Then, in order to make full use of the limestone, which was not completely reacted in the furnace, semidry emissions control technology based on dry FGD was carried out, and water was sprayed into the flue to further perform flue gas desulfurization. The flowchart before and after the upgrade of Hequ CFB Power Plant is as follows (Fig. 4.58).

(2) Coal quality

Hequ CFB Power Plant is a low calorific value coal comprehensive utilization power generation, all the coal burned was low calorific value bituminous coal. The coal quality in the whole year of 2016 was tested as Table 4.14 shown. At the same time, in order to discuss the change of coal quality more intuitively, the data was plotted and analyzed. The average value of the moisture content on as-received basis was 7.14%, and the moisture content on as-received basis does not exceed 1.5%. The ash content on as-received basis of coal-fired coals fluctuated greatly throughout the year, with a maximum of 47.86% and a minimum of only 34.16% with a large difference of more than 10%. The sulfur content on as-received basis was kept at 1% and the upper and lower floats did not exceed 0.1%. The average value of the volatile matter content on as-received basis was 22.76%, and the moving range was basically within 1% (Fig. 4.59).

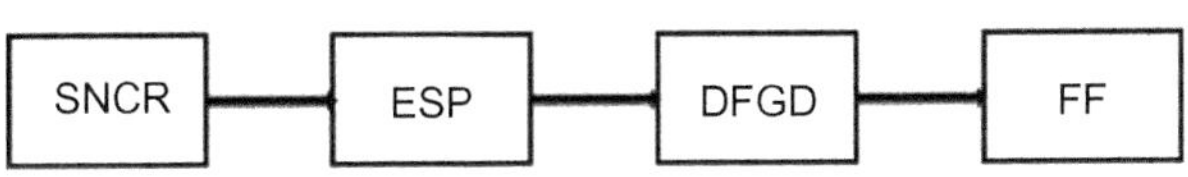

Fig. 4.58 Roadmap of Hequ CFB Power Plant.

Table 4.14 Coal quality analysis in Hequ CFB Power Plant

Month	Coal	Moisture (%)	Ash (%)	Sulfur (%)	Volatile matter (%)	Net calorific value (MJ/Kg)
1	Low-grade coal	5.94	40.85	1.21	23.15	14.83
2	Low-grade coal	5.60	36.71	1.24	23.79	16.40
3	Low-grade coal	6.09	36.41	1.00	23.62	16.30
4	Low-grade coal	7.56	34.16	0.98	24.12	16.20
5	Low-grade coal	8.58	34.28	1.01	23.81	16.17
6	Low-grade coal	8.28	34.19	1.01	23.79	16.34
7	Low-grade coal	7.93	38.67	0.90	22.23	14.77
8	Low-grade coal	7.39	39.83	0.96	22.34	14.52
9	Low-grade coal	7.16	42.87	0.88	22.75	14.63
10	Low-grade coal	8.01	44.23	0.82	21.28	12.56
11	Low-grade coal	6.94	45.05	0.97	21.57	12.81
12	Low-grade coal	6.23	47.86	0.89	20.68	11.95
Average		7.14	39.59	0.99	22.76	14.79

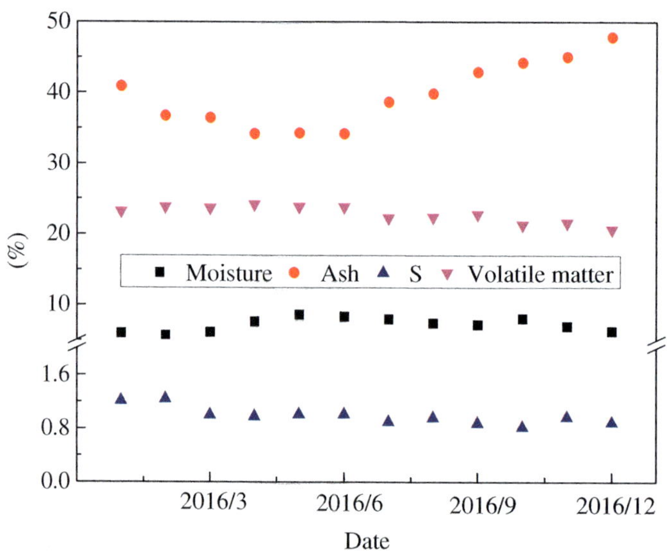

Fig. 4.59 Coal quality analysis in Hequ CFB Power Plant.

(3) Operation analysis of ultra-low emission

Figs. 4.60 and 4.61 show the characteristics of PM, SO_2, and NO_x emissions in the flue gas monitored hourly after the ultra-low emission implementation of Units 1 and 2 of the Hequ CFB Power Plant. The monitoring time was between November 1 to November 30, 2016.

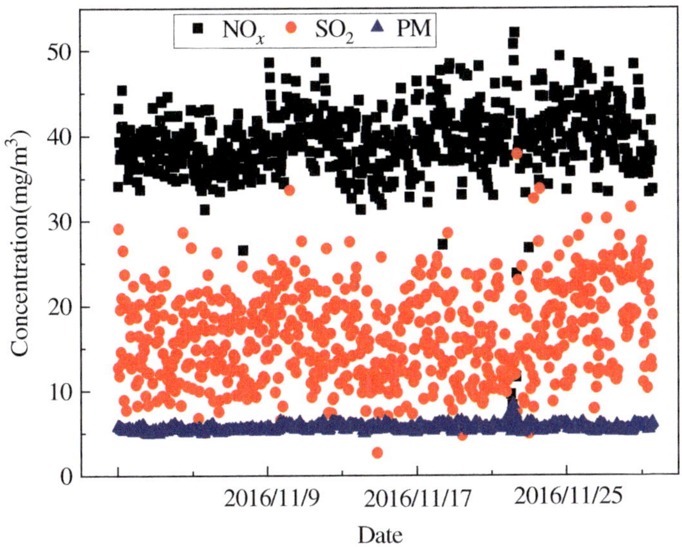

Fig. 4.60 Emission after ultra-low emission of Unit 1.

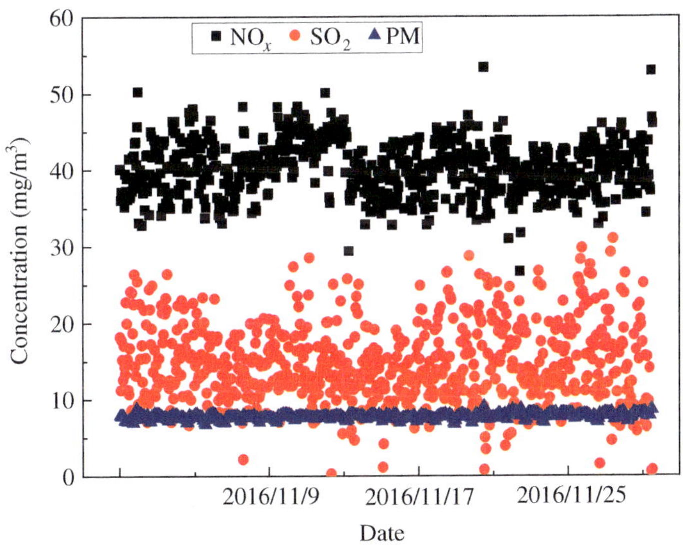

Fig. 4.61 Emission after ultra-low emission of Unit 2.

It showed that after the implementation of the ultra-low emission, the NO_x concentration of Unit 1of Hequ CFB Power Plant was maintained at 30 ~ 50 mg/m^3, the SO_2 concentration changed from 5 to 25 mg/m^3, and the PM concentration was also about 5 mg/m^3. After the

Table 4.15 Operation data after ultra-low emission of Unit 2

Emission	Annual average	Daily minimum	Daily maximum
PM	7.73	6.83	9.23
SO_2	16.54	2.33	30.85
NO_x	39.76	30.6	50.48

implementation of the ultra-low emission, the NO_x emission concentration of Unit 2 was less than 50 mg/m^3, and most of it was between 35 and 45 mg/m^3. Most of the SO_2 concentration ranged from 5 to 25 mg/m^3, and the PM concentration was below 10 mg/m^3, reaching ultra-low emission standards.

The analysis shows that the ultra-low emission technology adopted by the CFB unit had a good performance on the control of NO_x and SO_2. This indicated that the design of dry spray calcium and semidry method in the furnace had a good removal effect on SO_2 emissions from high-sulfur coal-fired units. Due to the high ash content of the coal itself, although the ultra-low emission requirements were also met, the PM concentration of the unit in Hequ Power Plant was relatively high compared to the previous ones. If the emission concentration was further reduced, it was necessary to comprehensively consider factors such as bag resistance.

Table 4.15 is the operating data of the unit in 2016 after the completion of ultra-low emission of Hequ CFB Power Plant unit. It showed that the average emission concentration of PM, SO_2, and NO_x was lower than the limit of the air pollutant emission concentration of the gas turbine unit, which basically met the ultra-low emission standard. The effect of the emission reduction was remarkable.

4.4.2 Midong Power Plant

(1) Introduction of the unit before upgrade

The Midong Thermal Power Plant was equipped with 2×300 MW subcritical condensing/extraction steam turbine units. It used DG1069/17.4-II1 boiler produced manufactured by Dongfang Boiler Co., Ltd. The boiler was a single steam drum, natural circulation, circulating fluidized bed combustion mode, and it was mainly composed of a membrane water-cooled wall, three cooling cyclone separators. ESP-FF was equipped with a PM removal efficiency of $\geq$99.91%. The denitrification part adopted the SNCR method with urea as the reducing agent, of which designed NO_x emission concentration was less than 100 mg/m^3. The SO_2 removal by two-stage desulfurization devices, involved first part was provided with a dry method with injection limestone in the furnace, and second-stage device of wet FGD was adopted outside the furnace. At last, the total desulfurization efficiency was more than 98%.

(2) Introduction to the upgrade roadmap

The upgrade for PM removal was considered from the following aspects: reducing the temperature of flue gas near the ESP, optimizing the existing PM collector, and adding new equipment. Addition of a low-temperature economizer can recover the waste heat of flue gas, and reduce the exhaust gas temperature to meet the original designed temperature and velocity condition. An integrated device of high-efficiency PM removal and demister was also equipped.

The desulfurization system was upgraded to meet the ultra-low emission requirement. The flue gas turbulence device was added and the demister was modified in the absorption system, to make the gas-liquid-solid fully contact and the mass transfer process to complete quickly. Thus, the purpose of gas purification was achieved, and the integration of desulfurization and demister was formed.

Based on the low bed temperature and the low bed pressure combustion characteristics of the circulating fluidized bed boiler, combustion adjustment, separator upgrade, secondary air upgrade, coal preparation system upgrade, SNCR denitrification system optimization, etc. were carried out to control NO_x in the units. The flow chart before and after the upgrade of Midong Power Plant is as follows (Fig. 4.62).

(3) Coal quality

Midong Power Plant is a low calorific value coal comprehensive utilization power generation, all the coal burned was mixed raw coal and coal gangue. The coal quality in the furnace from 2014 to 2016 was tested, and the specific coal quality is shown in Table 4.16. It showed that the average value of the moisture content on as-received basis was 17.5%, which belongs to the medium and high total moisture coal; the ash content on as-received basis was about 30%, which belongs to medium and high ash coal in the ash classification of thermal coal; the sulfur content on

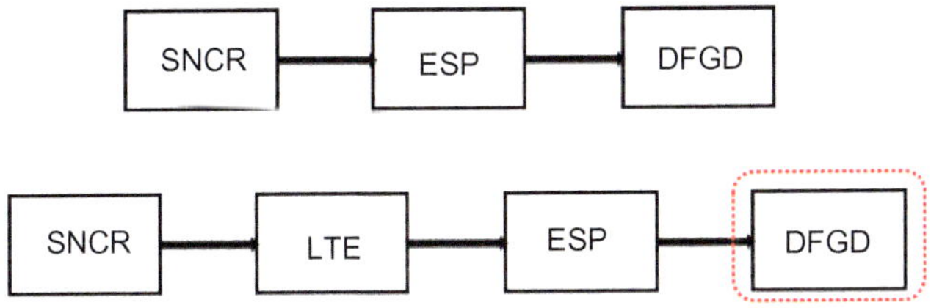

Fig. 4.62 Upgrade roadmap of Midong Power Plant

Table 4.16 Coal quality analysis in Midong Power Plant

Time	Coal	Moisture (%)	Ash (%)	Sulfur (%)	Volatile matter (%)	Net calorific value (MJ/Kg)
2014.1	Mixed coal	17.5	31.28	0.85	21.42	12.37
2014.2	Mixed coal	18.0	31.94	0.86	21.66	12.75
2014.3	Mixed coal	17.4	33.1	0.77	21.56	12.56
2014.4	Mixed coal	16.1	35.39	0.86	20.89	12.02
2014.5	Mixed coal	13.3	35.08	0.9	23.32	13.07
2014.6	Mixed coal	16.4	30.43	0.92	23.02	13.38
2014.7	Mixed coal	15.7	32.54	0.93	23.55	12.6
2014.8	Mixed coal	16.4	32.07	0.85	22.55	12.79
2014.9	Mixed coal	18.5	34.08	0.8	21.13	12.07
2014.10	Mixed coal	17.7	36.16	0.86	20.57	12.03
2014.11	Mixed coal	18.8	33.76	0.78	21.32	11.96
2014.12	Mixed coal	17.2	34.08	0.87	23.1	12.06
2015.1	Mixed coal	19.3	31.21	0.73	22.32	12.81
2015.2	Mixed coal	20.7	26.94	0.63	21.82	14.16
2015.3	Mixed coal	22.5	26.63	0.66	21.13	13.36
2015.4	Mixed coal	20.8	26.37	0.69	21.42	14.33
2015.5	Mixed coal	18.3	27.81	0.73	22.58	14.79
2015.6	Mixed coal	18.3	28.37	0.68	21.79	14.61
2015.7	Mixed coal	17.1	29.14	0.67	22.99	14.62
2015.8	Mixed coal	16.7	30.98	0.77	22.42	14.17
2015.9	Mixed coal	16.9	29.5	0.7	22.73	14.67
2015.10	Mixed coal	17.7	26.94	0.67	23.0	15.25
2015.11	Mixed coal	18.3	29.19	0.72	22.01	14.47
2015.12	Mixed coal	19.2	28.8	0.68	22.11	14.2
2016.1	Mixed coal	18.9	29.54	0.68	21.21	12.66
2016.2	Mixed coal	19.7	28.19	0.53	20.32	14.23
2016.3	Mixed coal	18.3	30.1	0.55	20.77	14.34
2016.4	Mixed coal	16.6	30.88	0.52	21.9	14.51
2016.5	Mixed coal	16.4	28.79	0.56	22.06	15.53
2016.6	Mixed coal	14.6	30.03	0.57	22.59	15.67
2016.7	Mixed coal	16.7	27.26	0.57	23.0	15.57
2016.8	Mixed coal	16	29.29	0.51	22.86	14.92
2016.9	Mixed coal	16.1	29.89	0.64	23.58	14.59
2016.10	Mixed coal	16	30.78	0.59	22.35	14.83
2016.11	Mixed coal	16.8	29.8	0.53	22.51	14.8
2016.12	Mixed coal	15.6	32.28	0.61	22.64	14.6
Average		17.5	30.51	0.70	22.11	13.82

as-received basis was about 0.7%, and it belongs to low sulfur coal in the sulfur classification; volatile matter content on as-received basis of about 22%, belonging to medium volatile coal; the base heat value was about 14MJ/Kg, which belongs to low calorific value coal.

(4) Operation analysis of ultra-low emission

Figs. 4.63 and 4.64 show the PM, SO_2, and NO_x emissions characteristics of the flue gas monitored hourly after the ultra-low emission upgrade of Unit 1 of the Midong Power Plant. The monitoring time was from January 1st to January 31st, 2016. Before the ultra-low

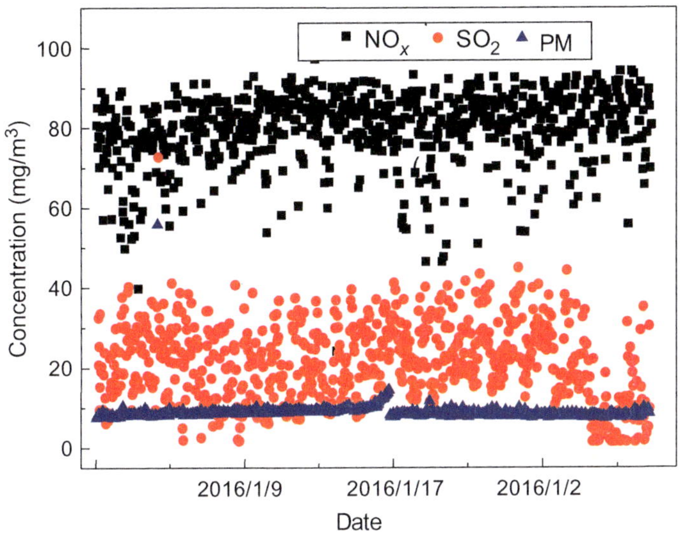

Fig. 4.63 Emission before ultra-low emission upgrade of Unit 1.

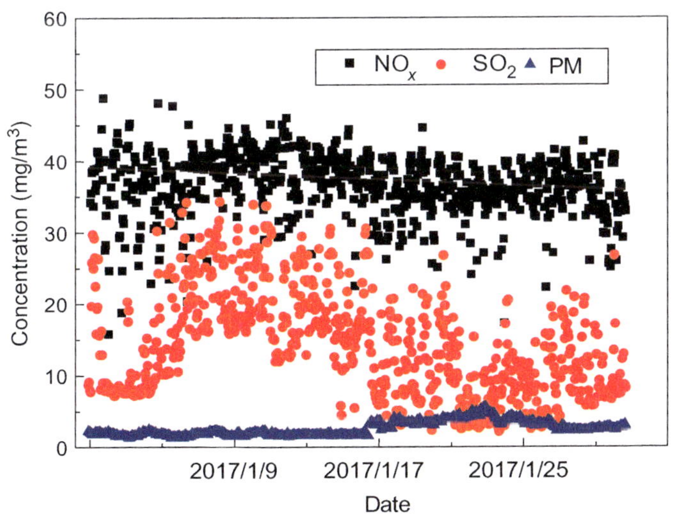

Fig. 4.64 Emission after ultra-low emission upgrade of Unit 1.

emission upgrade, the NO_x concentration of Unit 1 of Midong Power Plant was basically maintained at 50 ~ 90 mg/m^3, the SO_2 concentration varied between 5 and 40 mg/m^3, and the PM concentration was between 5 and 10 mg/m^3. After the ultra-low emission upgrade, the NO_x emission concentration of Unit 1 was less than 50 mg/m^3, and most of them were between 30 and 40 mg/m^3; most of the SO_2 concentration ranged from 5 to 30 mg/m^3, and the PM concentration fell below 5 mg/m^3, meeting the ultra-low emission requirements.

Comparing the changes of pollutant emission concentration of CFB units before and after the upgrade showed that the NO_x and the PM emission concentration had a significant decrease, indicating that the synergistic effect of the low-temperature economizer and the desulfurization system demister was satisfactory; optimization of denitrification in the furnace had an obvious effect on NO_x removal.

Table 4.17 shows the running data in 2017 of the unit after the completion of the ultra-low emission upgrade of the unit of the Midong Power Plant. All in all, after ultra-low emission conversion, the concentration of PM, SO_2, and NO_x emissions in the long-term was lower than 5, 40, 50 mg/m^3, in line with ultra-low emission standards. The average emission concentration of PM, SO_2, and NO_x after the upgrade of the Midong Power Plant Unit was lower than the air pollutant concentration limit of the gas turbine unit, which met the ultra-low emission standard. Significant reduction in emissions was achieved through the upgrade.

In most cases, the coal quality of CFB boiler was in the low rank coal condition, generally with high ash and sulfur content. For the unique characteristics of the CFB boiler combustion and coal quality, it has a challenge to meet ultra-low emission requirements. Among these four technology roadmaps, from the data of these cases, after ultra-low

Table 4.17 Operation data of Unit 1 in Midong after ultra-low emissions upgrade

	PM (mg/m^3)			NO_x (mg/m^3)			SO_2 (mg/m^3)		
Month	Average	Daily min	Daily max	Average	Daily min	Daily max	Average	Daily min	Daily max
1	2.5	2.3	2.9	38.7	35.5	41.9	13.8	12.8	14.8
2	2.8	2.5	3.4	38.3	35.2	42.8	15.9	15.1	16.1
3	3.9	2.5	4.1	36.4	33.9	48.7	15.6	14.9	16.7
4	2.4	2.4	4.2	37.9	36.5	40.8	21.3	18.7	18.4
5	2.8	2.6	4	31.8	30.9	41.9	10.2	10.1	17.1

emission implementation, the pollutant concentration of the CFB could reach the ultra-low emission requirement on the existing basis, but whether the emission concentration could be further reduced needs to be further studied.

4.5 Emissions of pollutants under different loads

In actual running operation, the change of loads had a certain impact on pollutant emissions, especially the SCR catalyst deviated from the work temperature window range under low load, and the NO_x emission had a chance to be very high. This section discussed pollutant emissions under different loads of typical units.

4.5.1 Hequ Power Plant

Figs. 4.65, 4.66, 4.67, and 4.68 show the change of pollutant concentration with loads before and after the ultra-low emission upgrade of Hequ Power Plant. It was found that the concentration of PM did not change significantly with the load. The NO_x concentration decreased significantly after the ultra-low emissions upgrade compared to before the upgrade. However, except for the NO_x emission of Unit 4, which had a certain downward trend with the increase of load, the overall emission did not change significantly

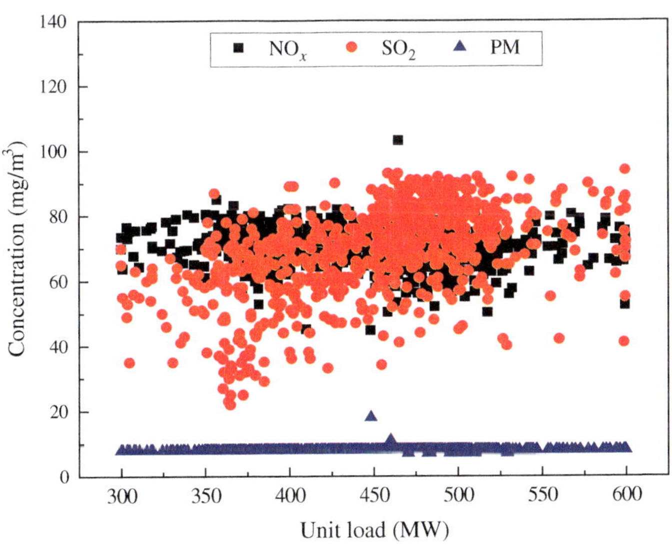

Fig. 4.65 Emission before ultra-low emission upgrade of Unit 3.

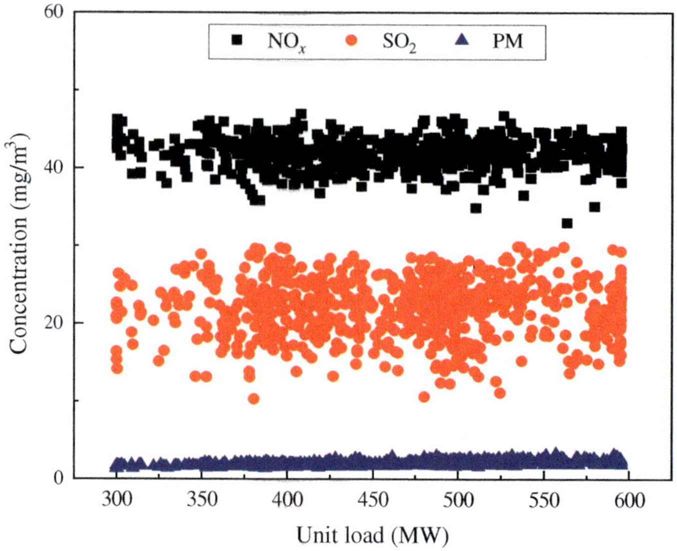

Fig. 4.66 Emission after ultra-low emission upgrade of Unit 3.

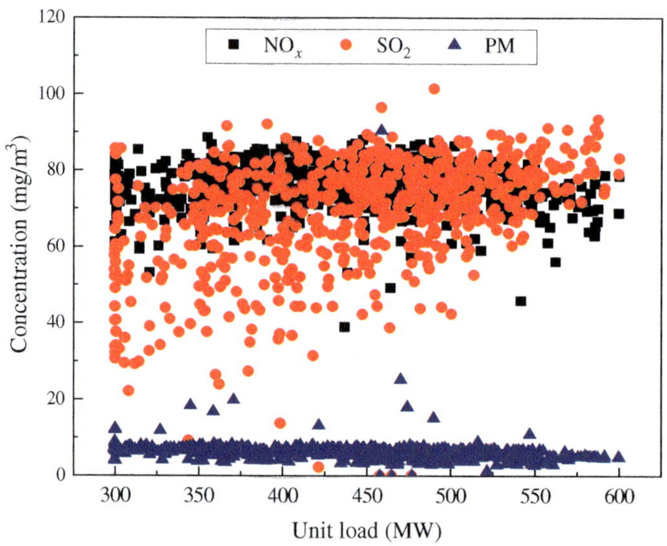

Fig. 4.67 Emission before ultra-low emission upgrade of Unit 4.

with the load. This may be attributed to two reasons: on the one hand, the increase in load led to an increase in the amount of NO_x generated by the increase in furnace temperature, on the other hand, the increase in the load made the catalytic performance of the SCR catalyst better, which resulted in a small change in the emission concentration. The SO_2 concentration

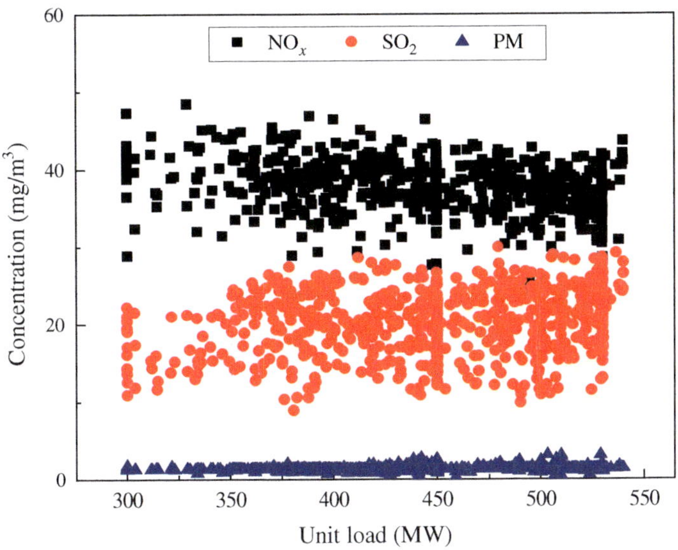

Fig. 4.68 Emission after ultra-low emission upgrade of Unit 4.

decreased significantly before and after the upgrade, but unlike the previous two pollutants, the SO_2 emission concentration showed a positive correlation with the unit load. This was the result of an increase in the amount of coal burned and an increase in the amount of SO_2 produced as the unit load increased.

4.5.2 Wanzhou Power Plant

Figs. 4.69 and 4.70 show the characteristics of PM, SO_2, and NO_x emissions from Units 1 and 2 under different loads after ultra-low emission upgrade in Wanzhou Power Plant. The PM, SO_2, and NO_x emission concentrations in different load ranges were lower than 5, 15, and 25 mg/m^3. As the load increased during running operation, the PM was stable within a certain concentration range and had a slight increase trend. SO_2 had a certain decreasing trend, while NO_x fluctuated greatly with the change of load, but the overall had a certain increasing trend. During the actual operation of the unit, as the load increased, the amount of coal burned increased and the total amount of pollutants increased. Although the treatment of high-pressure electrostatic precipitators and tube-type PM collector devices had a role in PM collector and emission reduction was the same as the load, the comprehensive PM collector efficiency was already high. Therefore, the concentration of PM emissions had only a small upward trend. The emission concentration of SO_2 showed a certain decreasing trend, which was different from the

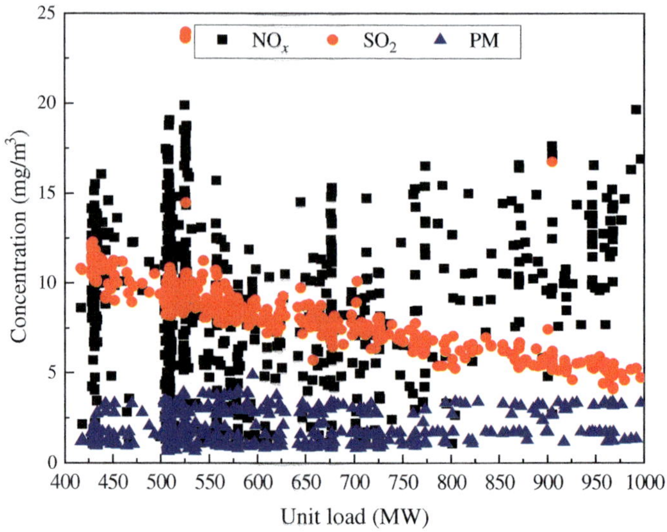

Fig. 4.69 Emission after ultra-low emission of Unit 1.

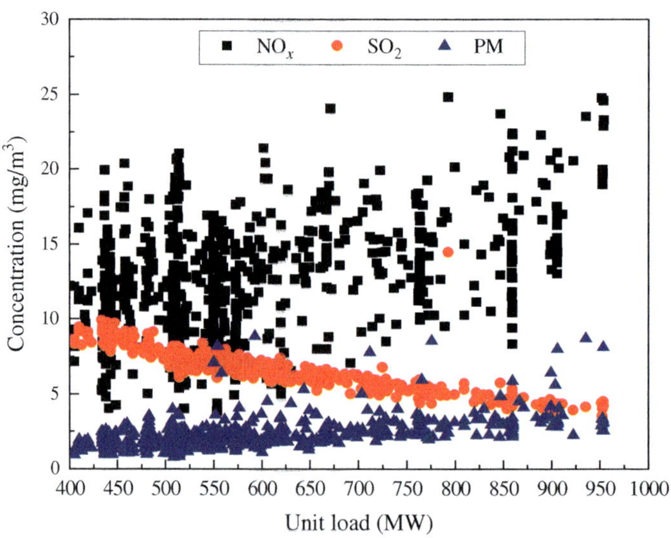

Fig. 4.70 Emission after ultra-low emission of Unit 2.

afore-mentioned Hequ Power Plant. As the load increased, the limestone amount was increased during operation, and the desulfurization efficiency was increased, so that the emission concentration was also decreased. For NO_x, the SCR catalyst had a low catalytic oxidation conversion efficiency in low-temperature conditions during low load. As the load increased, the temperature of the flue gas rose, and the reaction could occur under a suitable temperature window, thereby reducing the NO_x emission concentration. This could offset the increase in NO_x concentration caused by the increase in fuel caused by the increase in load. In general, the effect of load increase was greater than the effect of temperature on the SCR catalyst, resulting in a general upward trend in NO_x.

4.5.3 Qinhuangdao Power Plant

Figs. 4.71–4.76 show the characteristics of PM, SO_2, and NO_x emissions from Unit 1/2/3 under different loads of the ultra-low emission upgrade in Qinhuangdao Power Plant. As the load increased, the PM was stable within a certain range of emission concentrations and had a small increase trend. SO_2 had a certain decreasing trend, and NO_x fluctuated greatly with the change of load, but the overall had a certain increasing trend, and the result was similar to Wanzhou Power Plant.

4.6 Characteristics of emission and distribution at different locations

Different APCDs had different influences on PM, SO_2, and NO_x pollutants. It was important to understand the performance of this equipment by monitoring relevant pollutants before and after them. Limited by the data

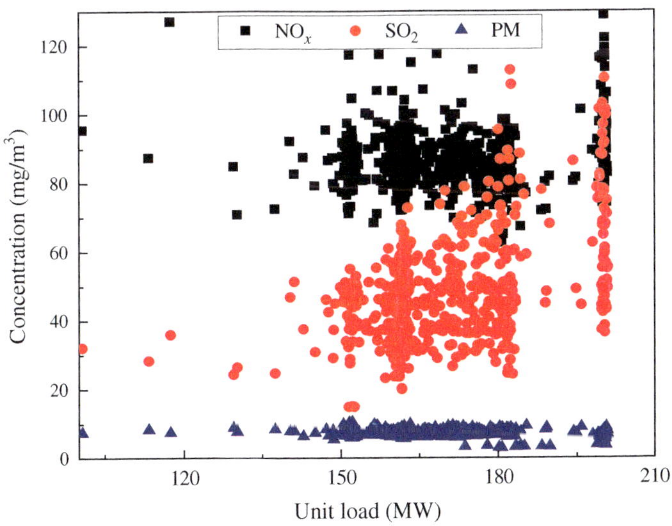

Fig. 4.71 Emission before ultra-low emission upgrade of Unit 1.

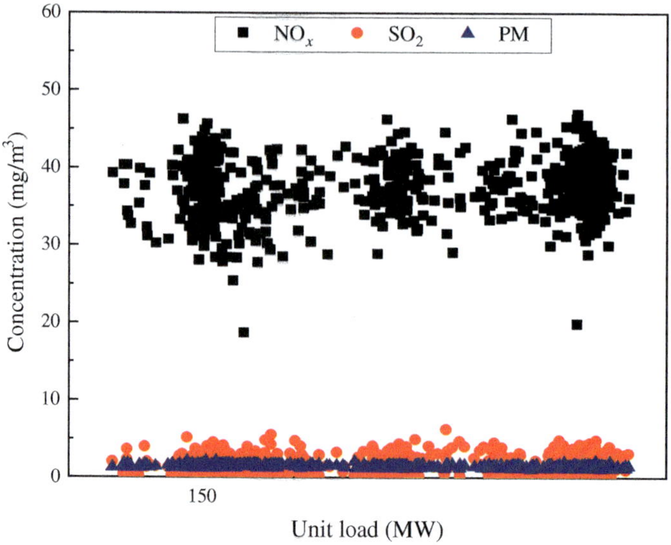

Fig. 4.72 Emission after ultra-low emission upgrade of Unit 1.

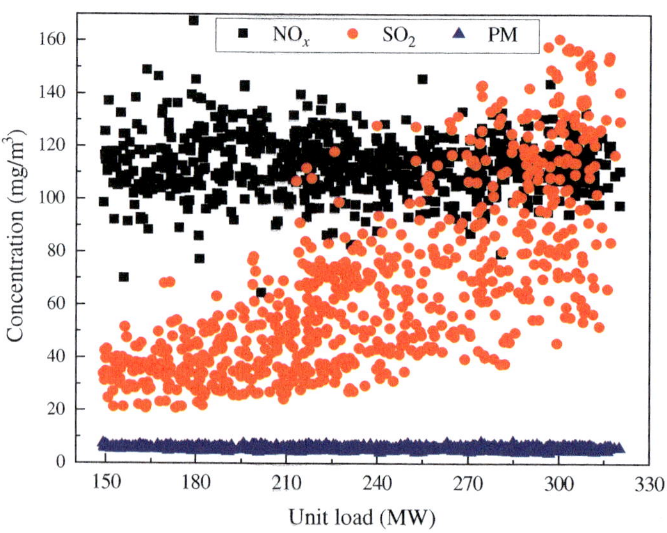

Fig. 4.73 Emission before ultra-low emission upgrade of Unit 3.

acquisition conditions, this section only took the Unit 3 of Dagang Power Plant as an example to discuss the emissions and changes of various APCDs.

NO_x emission before and after SCR, PM, SO_2, and NO_x emission after FGD were analyzed by online monitoring data after ultra-low emission upgrade. In Fig. 4.77, PM emission concentration showed that after the

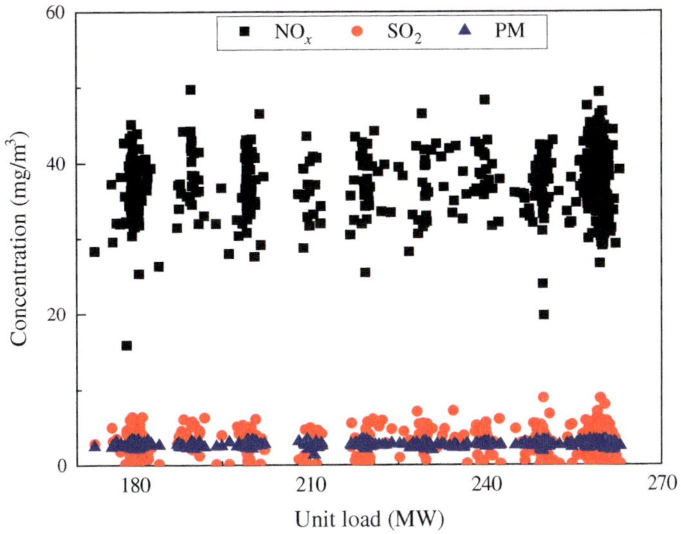

Fig. 4.74 Emission after ultra-low emission upgrade of Unit 3.

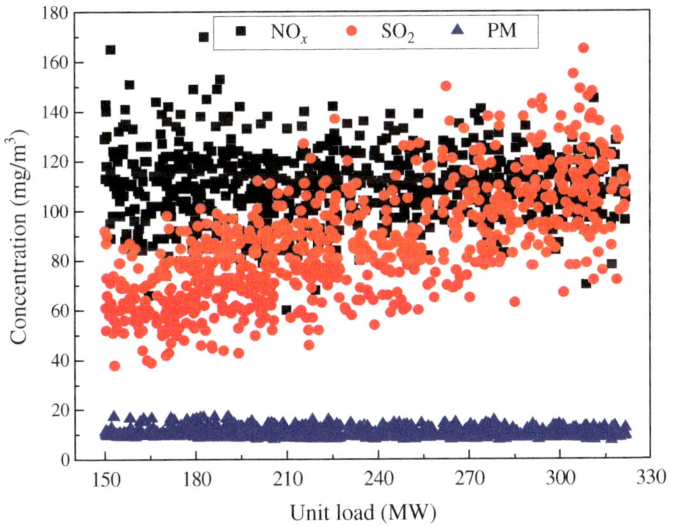

Fig. 4.75 Emission before ultra-low emission upgrade of Unit 4.

combined action of the foregoing low-temperature economizer and the three-phase power source modified ESP, the PM concentration in FGD inlet was about 5 mg/m^3, and the PM concentration in the stack inlet (FGD outlet) was only about 2.5 mg/m^3. The removal efficiency of FGD on PM was about 50%. Desulfurization upgrade was mainly through the

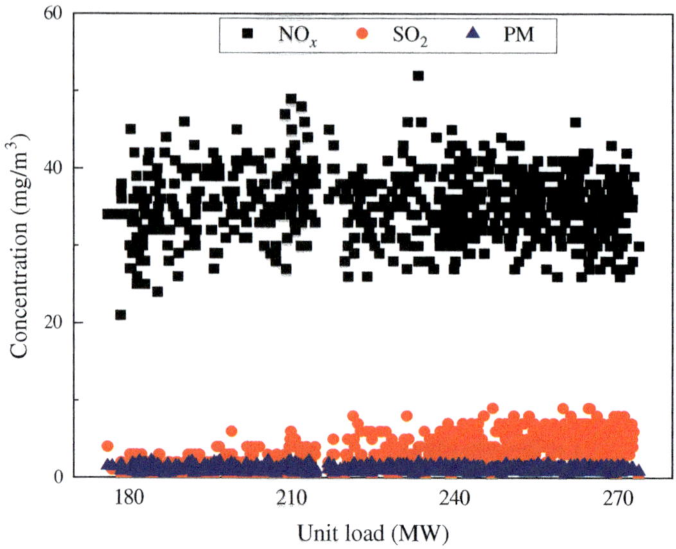

Fig. 4.76 Emission after ultra-low emission upgrade of Unit 4.

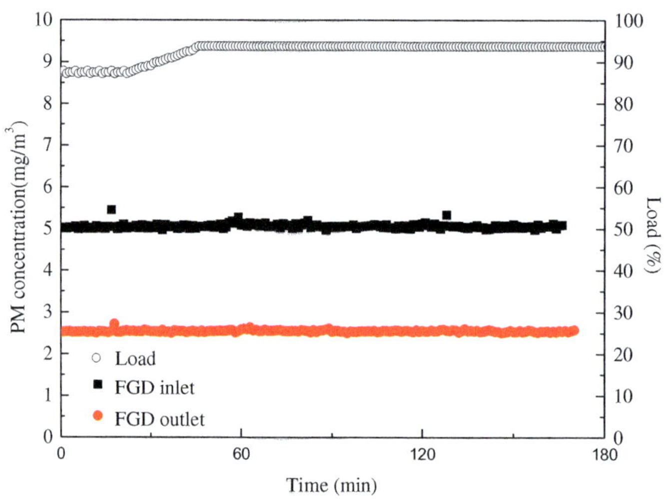

Fig. 4.77 Characteristics of PM concentration in the upgrade Unit 3 (upper is day 1, below is day 2).

additional differentiated arrangement of high-efficiency three-story roof ridge demister in FGD system. The online monitoring data before and after FGD (Fig. 4.78) showed that the concentration of SO_2 dropped sharply from 950 to about 20 mg/m^3, and the desulfurization efficiency of FGD

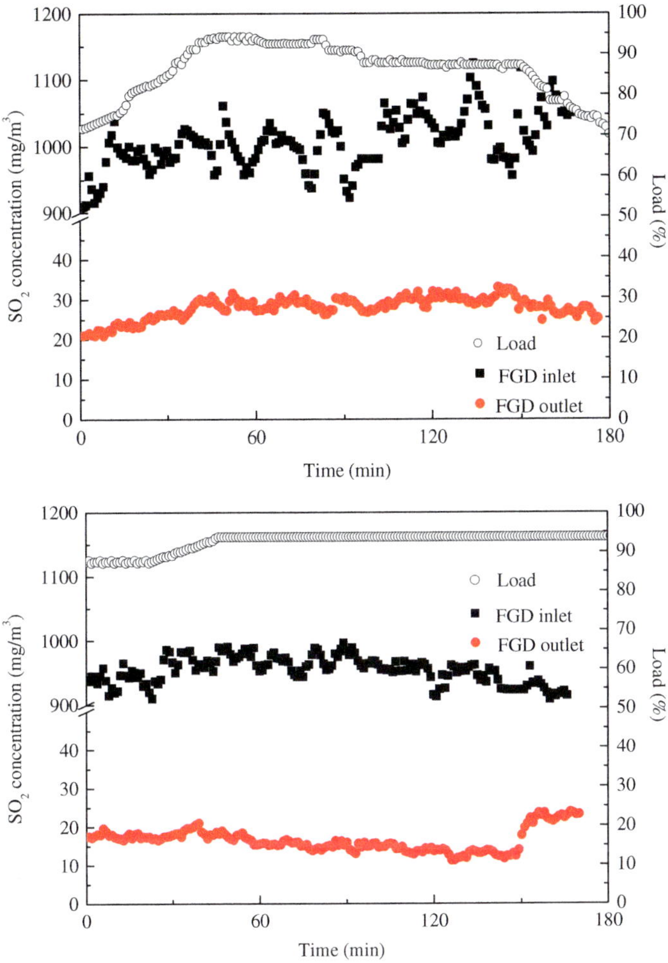

Fig. 4.78 Characteristics of SO_2 concentration in the upgrade Unit 3 (upper is day 1, below is day 2).

could reach 97.9%. NO_x emission reduction was realized through low nitrogen burners and SCR catalysis. The unit was equipped with NO_x online monitoring instrument at the SCR inlet, outlet, and FGD outlet. The results showed (Fig. 4.79) that after low nitrogen burners, the concentration of NO_x was about 300 mg/m^3, and after SCR, the concentration significantly decreased to about 30 mg/m^3, and the denitrification efficiency of SCR could reach more than 90%. At the same time, from the data of the monitoring period, there was no significant difference between pollutant emission and load change.

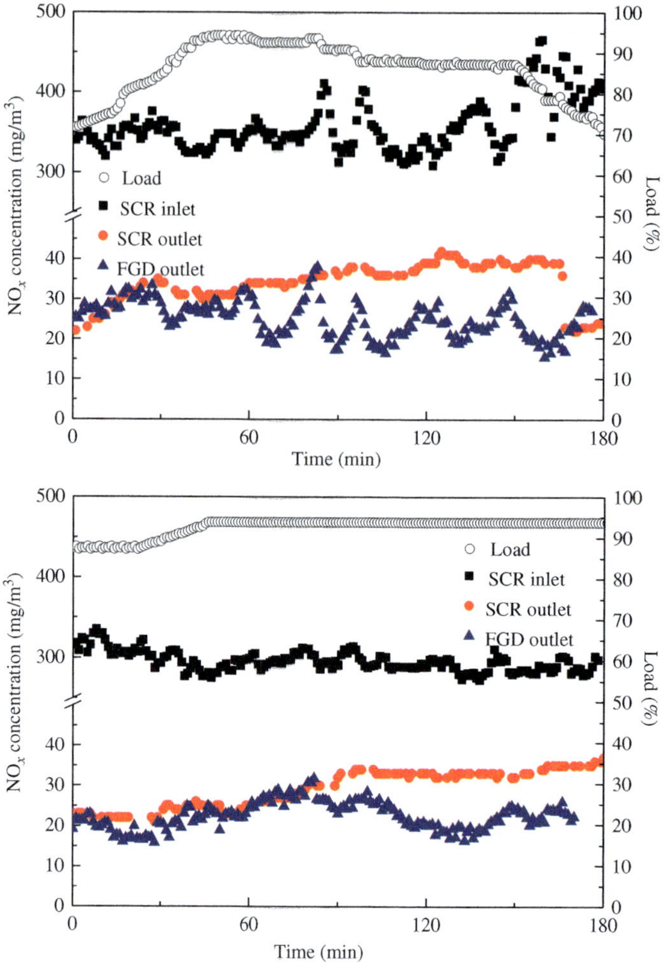

Fig. 4.79 Characteristics of PM concentration in the upgrade Unit 3 (upper is day 1, below is day 2).

4.7 Summary

The implementation of ultra-low emissions upgrade was goal-oriented, and the overall goal was to meet emission requirements through the application of relevant technologies. Before the implementation of ultra-low emission, the unit construction, coal quality, and APCDs were very different for different power plant. The emission limits of 10, 35, and 50 mg/m^3 needed to be based on the current technical level and equipment conditions. From the perspective of the development of the coal-fired power plants, the technology with low cost, mature technological level, and reliable operation could be the best choice. In this chapter, more than 20 units were related to the implementation of four emissions control technologies, including Emissions Control Technology Based on DESP and WESP, Emissions Control Technology Based on LTT-ESP, Emissions Control Technology Based on Hybrid ESP-FF, and Emissions Control Technology Based on Dry FGD (DFGS) were investigated, and the main results were summarized in Table 4.18.

The pollutant emission level and emission characteristics were investigated under different loads, before and after upgrade and long period operation. The main conclusions were as follows:

(1) For different pulverized coal boilers and circulating fluidized bed boilers with bituminous, lignite and low rank coal, based on four different upgrade roadmaps, the emission concentrations of PM, SO_2, and NO_x after the ultra-low emission upgrade could be lower than 10, 35, and 50 mg/m^3, respectively. Based on the running condition of the unit, the applicable scopes of the four technology roadmaps were analyzed. For the future ultra-low emission upgrade, it is necessary to implement the technical scheme of flue gas pollutant control technology combined with the specific characteristics of units and coal quality, and the performance needs further investigation for these applications.

(2) Through the analysis of the running operation data of the Hequ Power Plant, Wanzhou Power Plant, and Qinhuangdao Power Plant, it was found that the operational load had a certain impact on pollutant emissions, but the pollutant emission concentration after the implementation of the ultra-low emission upgrade could reach the ultra-low emission standard.

(3) The emission at different locations of Dagang Power Plant was monitored. The NO_x concentration before SCR was 300 mg/m^3, and the NO_x concentration fluctuated between 20 and 40 mg/m^3 after FGD. The concentration of SO_2 dropped from 950 to about 20 mg/m^3. The PM in the FGD inlet was about 5 mg/m^3, and the PM in the stack inlet (FGD outlet) was only about 2.5 mg/m^3.

Table 4.18 Comparison on technology roadmaps and implementation effects of different units

	Technology roadmaps	Power plant	Units	Capacity (MW)	Technical features	Coal quality			Emission reduction: PM (mg/m^3)			SO_2 (mg/m^3)			NO_x (mg/m^3)		
						Classification	Ash (%)	Sulfur (%)	Before-upgrading	After upgrading	Emission reduction (%)	Before - upgrading	After upgrading	Emission reduction (%)	Before-upgrading	After upgrading	Emission reduction (%)
1	DESP, WESP	Wnagqu	#1	600	An extra layer of catalyst added, three-phase power supply, roof ridge -style demister	Lean & bituminous	19.69	0.34	2.19	0.88	59.8	73.66	16.38	77.8	82.89	38.73	53.3
2		Wanzhou	#1	1050	LNB, SCR, high voltage/high-frequency/ three-phase power supply, spin-coupled WFGD	Bituminous	10.69	0.39	Newly build	2.48	/	Newly build	9.54	/	Newly build	15.36	/
3			#2	1050					Newly build	2.29	/	Newly build	7.93	/	Newly build	12.81	/
4		Zhoushan	#4	350	LNB, SCR, ESP, sea water desulfurization, WESP				Newly build	2.33	/	Newly build	3.71	/	Newly build	22.19	/
5	LTT-ESP	Yuanyanghu	#1	660	Combustion adjusting, LTE, three-phase high-frequency power supply, efficient demisting	Bituminous	23.95	0.85	28.8	3.26	88.7	97.2	21.20	78.2	75.8	38.14	49.7
6			#2	660					26.8	2.89	89.2	105.4	21.11	80.0	66.1	42.03	36.4
7		Hequ	#3	600	A extra layer of catalyst added, LTE, FF removal changed into ESP, efficient demisting	Lignite	20.21	0.57	8.00	2.06	74.3	69.27	22.14	68.0	70.06	41.83	40.3
8			#4	600					10.94	1.43	86.9	70.88	20.35	71.3	77.43	37.72	51.3
9		Dagang	#1	329	LNB, LTE, three-phase power supply, three-stage roof ridge-style demisting	Bituminous	17.11	0.43	7.32	3.3	54.9	19.24	16.13	16.2	34.71	24.8	28.6
10			#2	329					5.6	3.11	44.5	19.01	16.53	13.0	33.46	30.44	9.02
11			#3	329					2.89	2.47	14.5	16.55	16.11	2.65	48.35	30.07	37.8
12			#4	329					4.94	3.03	38.7	16.24	15.77	2.8	27.62	22.87	17.2

13		Sanhe	#1	350	LNB, SCR, LTE, high-frequency ESP, roof ridge-type tube demisting, WESP	Bituminous			/	2.72	/	/	16.6	/	/	30.77	/
14			#2	350					/	1.82	/	/	15.22	/	/	29.74	/
17		(Zhejiang)	#7	1000	LNB, An extra layer of catalyst added, LTE, high-frequency ESP, double-layer pallet for WFGD, WESP	Bituminous	11.73	0.48	/	5.2	/		8.6	/	/	26.3	/
19	ESP-FF	Jiaozuo	#1	660	Enhanced ammonia spray, FF+ESP, high-frequency power supply, efficient demisting	Lignite	19.33	0.64	7.75	3.73	51.9	24.91	12.16	51.2	31.51	24.35	22.7
20			#2	660					6.46	2.84	56.0	48.28	14.62	69.7	37.69	19.39	48.6
21		Qinhuangdao	#1	215	A extra layer of catalyst added, FF+ESP, seawater desulfurization, efficient demisting	Bituminous	21.89	0.58	7.31	1.44	80.3	47.91	1.69	96.5	86.62	37.53	56.7
22			#3	320					5.68	2.75	51.6	66.48	3.00	95.5	113.16	36.99	67.3
23			#4	320					10.71	1.18	89.0	86.07	3.32	96.1	110.07	35.28	67.9
24		Huayuan	#3	660	An extra layer of catalyst added, FF+ESP, enhanced SCR, efficient demisting	Lignite	21.29	0.29	13.45	2.72	79.8	42.15	16.18	61.6	27.56	11.95	56.6
25	DFGD	Hequ CFB	#1	350	Calcium spraying inside furnace, SNCR, FF+ESP, semidry desulfurization	Low grade	39.59	0.99	Newly build	5.64	/	Newly build	16.51	/	Newly build	39.24	/
26			#2	350					Newly build	7.79	/	Newly build	14.69	/	Newly build	40.28	/
27		Midong CFB	#1	300	Calcium spraying inside furnace, SNCR, LTE, WFGD, FF after desulfurization	Low grade	30.51	0.70	7.48	2.88	61.5	20.58	15.36	25.4	78.58	36.62	53.4

Reference

[1] S. Wei, L. Li, Z.M. Cui, J.Y. Wu, H. Mo, Analysis of primary air pollutant emission characteristics and reduction efficiency for ultra-low emission coal-fired power plants based on actual measurement, Electric Power 48 (11) (2015) 131–137.

CHAPTER FIVE

Fine particles characteristics of ultra-low emission coal-fired power plants

Yongsheng Zhang
School of Energy, Power and Mechanical Engineering, North China Electric Power University, Beijing, China

5.1 Introduction

Atmospheric particulate matter can be classified according to aerodynamic diameter, as shown in Fig. 5.1. Particulate matter, whose size is less than 10 μm, (PM_{10}) is also known as inhalable particulate matter because it can enter and deposit in the human body. In recent years, the phenomenon of haze in China has become more and more frequent, which is mainly caused by the increase of PM_{10}, especially the particulate matter with aerodynamic diameter less than 2.5 μm ($PM_{2.5}$) [1]. In the 2012, Environmental Air Quality Standard (GB3095-2012) released by China, $PM_{2.5}$ has become an important new monitoring indicator.

PM_{10} has complex chemical components, most of which contain inorganic compounds such as sodium nitride, calcium fluoride, barium compounds, iron, vanadium compounds, and some of them also carry benzene and homologs [2]. At the same time, because $PM_{2.5}$ could stay in the atmosphere for quite a long time, it is easy to enrich the toxic heavy metals, acid oxides, organic pollutants, bacteria, and viruses in the air [3], and has become the carrier to make these substances have higher reaction and dissolution rate. $PM_{2.5}$ is considered to have more health impacts than the relatively large particles. Exposure to fine particles can affect lung function and worsen respiratory medical conditions. People with breathing and heart problems, children, and elderly people are particularly sensitive to $PM_{2.5}$. Particulate matter is linked to lung cancer, chronic obstructive pulmonary disease, and heart diseases. $PM_{2.5}$, whether inhaled indoors or outdoors, are particularly harmful to health

Advances in Ultra-low Emission Control Technologies for Coal-Fired Power Plants
https://doi.org/10.1016/B978-0-08-102418-8.00005-X

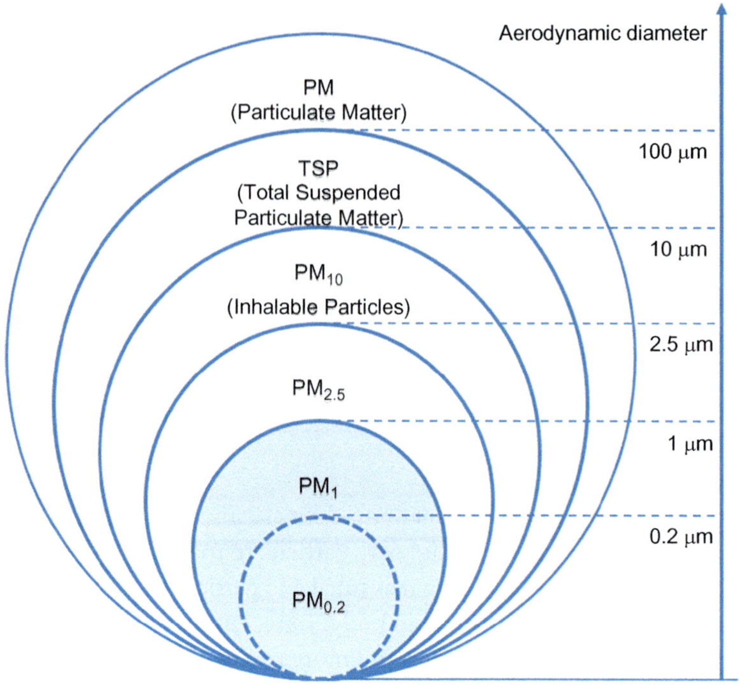

Fig. 5.1 Particulate matter classification.

as they can penetrate deep into the lungs [4]. In particular, $PM_{2.5}$ and undissolved contaminants could enter and deposit in the lungs and has a possibility of entering the blood. The increase of $PM_{2.5}$ is closely related to the morbidity and mortality of the human respiratory or cardiovascular system [5, 6].

As one of the important sources of PM_{10}/$PM_{2.5}$, the emission of PM_{10}/$PM_{2.5}$ from coal-fired power plant has also received more attention. Most of the conventional coal-fired power plants use high-efficiency PM removal technologies such as electrostatic precipitator, fabric filter, etc. Although the particles can be effectively removed, the removal effect of PM_{10}/$PM_{2.5}$ is not ideal compared with large particles. In particular, the control effect of $PM_{2.5}$ is not obvious while it is dominant in number concentration. How to improve the removal effect of ultra-low emission power plant upgrading on fine particles has become one of the challenging targets for power companies, design companies, and research institutes.

As described in Chapter 3, the ultra-low emissions of particulate matter from coal-fired power plants are based on two major upgrades:

5.1.1 Upgrade of primary PM removal technology

Primary PM removal technology (control of total PM in flue gas after SCR) is the main PM control equipment for coal-fired power plants. Main technologies include electrostatic precipitator, fabric filter, and electric fabric filter composite precipitator. The combination of high-efficiency power source upgrading, cleaning method optimization, and low-temperature electric PM removal technology for electrostatic precipitator technology can improve the PM removal efficiency and improve the removal efficiency of fine particles.

5.1.2 Development of deeply removal technology for PM after desulfurization

These technologies are known as secondary PM removal/deep PM removal techniques; they include the upgrading of the defogging system of the wet desulfurization system or the installation of the wet electrostatic precipitator (WESP), the technical route for the dry or semidry desulfurization bagging PM collector and the electric fabric filter composite PM collector. The application of these technologies has reached the standard of ultra-low emissions while improving the control efficiency of fine particles.

In general, the mechanism and actual cases of ESP, electric fabric filter composite PM collector, FGD and WESP on fine particles will be discussed in this chapter. Since there is no online monitoring equipment for PM_{10} and $PM_{2.5}$ in coal-fired power plants at present, the experiments in this chapter mainly use Electrical Low-Pressure Impactor (ELPI) and Gravimetric Impactor for field test.

5.2 Impact of ESP on $PM_{2.5}$

5.2.1 The mechanism of upgraded ESP on $PM_{2.5}$

Electrostatic precipitator technology is the preferred PM removal technology in the power industry because of its advantages of large amount of flue gas, high PM removal efficiency, low equipment resistance, simple and reliable use, low operation and maintenance cost, and no secondary pollution [7–10]. In the model selection design stage, traditional electrostatic precipitators are mainly based on coal and ash characteristics, most of them do not have the ideal removal effect on $PM_{2.5}$ [7, 11]. The removal of particulate matter larger than 10 μm in an electrostatic precipitator is mainly determined by electrostatic force and Stocks resistance, while $PM_{2.5}$ is mainly related to charge and gas flow distribution [12]. The charging and airflow distribution are mainly affected by factors such as power source technology, electrostatic precipitator body arrangement, and flue gas temperature. Conventional ESP has problems such as backward power source technology, insufficient design targeting, and high operating temperature, which makes it less effective for $PM_{2.5}$ control. In response to these problems, ultra-low emission power plants have made some upgrading to the ESP system, thereby increasing its control effect on $PM_{2.5}$.

The process of trapping particulate matter by electrostatic precipitator includes four steps: (1) negative high-voltage discharge, flue gas ionization, generation of negative ions; (2) fly ash is charged by electric field and negative ions; (3) charged fly ash is collected to the anode plate; (4) after rapping, the fly ash on the anode plate falls into the hopper [12, 13]. At present, the electric PM removal design of the power industry is mainly based on the Deutsch correction formula (Eq. 5.1) as Chapter 3 mentioned [14]:

$$\eta = 1 - e^{-\left(\omega_k \cdot \frac{A}{Q}\right)^k}, \tag{5.1}$$

in which ω_k is the effective driving speed of fly ash, cm/s; A is the total fly ash collecting anode plate area of electric PM removal, m^2; Q is the flue gas flow to be treated by electrostatic precipitator, m^3/s; k is the empirical coefficient, generally 0.5. The specific collection area (SCA) of ESP is equal to A/Q. When the flue gas volume is constant, the electric PM removal efficiency can only be improved by increasing the area of the PM collection plate.

However, due to the large deviation between actual value and the PM model selection designed according to driving speed, the traditional PM model selection method has been difficult to deal with the 10–30 mg/m^3 particulate matter emission and to give a reasonable design and upgrading plan [12]. Moreover, the electrostatic precipitating efficiency depends to a large extent on the secondary voltage and current of the precipitator, and the cathode or anode PM collection, electrode deformation or displacement, etc. may cause a great change on the PM removal efficiency. According to formula (5.1), real-time PM removal efficiency prediction is difficult to achieve [12].

The efficiency of electrostatic precipitator depends mainly on the size of the body and the operating voltage. On the basis of the PM removal efficiency and total PM removal efficiency of the electrostatic precipitator,

Yan Keping team of Zhejiang University proposed to use the electric PM removal index for the selection design [12, 15]:

$$\lg\frac{1-\eta(r)}{\beta} = -\alpha \cdot E_a \cdot E_p \cdot S, \tag{5.2}$$

where $\eta(r)$ is the graded PM removal efficiency; $E_a \cdot E_p \cdot S$ is defined as the electric PM removal index, which represents the electric field energy density obtained by the flue gas volume in the unit PM collection area; α is the operation condition coefficient of the electrostatic precipitator, $(\text{cm/kV})^2$ m/s; β is the system operating condition coefficient; E_a is the average electric field strength of the electrostatic precipitator, kV/cm; E_p is the peak electric field strength of the electric precipitator, kV/cm.

Fig. 5.2 shows the Yan Keping study on the relationship between the electrostatic precipitator index and the fine particle removal efficiency. By adjusting the voltage or flue gas volume to change the electrostatic precipitator index, as the electrostatic precipitator index increases, the PM collection efficiency also increases. When the electrostatic precipitator index reaches 200 $(\text{kV/cm})^2\ \text{m}^2/\text{m}^3/\text{s}$, the collection efficiency is over 95%. In particular, the removal efficiency of the particles of 0.1–1 μm is improved significantly [16]. Most of the theoretical analysis about the effects of electrostatic precipitator upgrading on fine particulate matter in this chapter focuses on the electrostatic precipitator index.

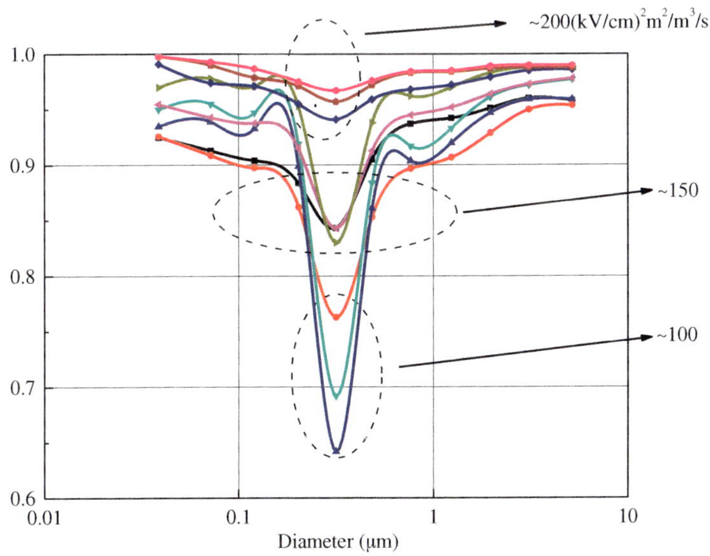

Fig. 5.2 ESP graded collection efficiency. *(Adapted from K.P. Yan, Electric precipitation upgrading of coal-fired power plant: basic principle and industrial application, in: Proceedings of the 17th Conference of Electrostatic Precipitation, Chinese Society of Electrostatic Precipitation, 2017.)*

5.2.2 Different power sources

The power source is an important part of the ESP. In the ultra-low emission upgrading, some units use high-voltage power sources such as high-frequency power sources or three-phase power sources instead of traditional single-phase power sources, and some ESPs have been upgraded with pulsed power sources. The difference between different high-voltage power sources is operating voltage, current (secondary voltage, current), peak voltage, and power factor [17].

The traditional ESP mainly uses single-phase power frequency power source, and its ripple coefficient (full load) ranges from 50% to 100%, while the three-phase power source is less than 5%, and the high-frequency power source is less than 1% [17]. Therefore, under the same power source capacity condition, the highest secondary voltages of the precipitator of the three-phase power source and the high-frequency power source are basically equal, and both of them are higher than that of the single-phase power source. Additionally, the power conversion coefficients of the three-phase power source and the high-frequency power source are also higher than the single-phase power source. Therefore, under the same output power conditions, the power utilization of the three-phase power source and the high-frequency power source are higher than those of the single-phase power source. At the same time, the three-phase power source can increase the migration rate of particulate matter from 17 to 25–35 cm/s [18].

Table 5.1 shows the performance comparison of single-phase power, three-phase power, and high-frequency power. The power factor of the

Table 5.1 Power sources performance [17]

Parameter	Single phase	Three phase	High frequency
Design average voltage (kV)	72	80	85
Design peak voltage (kV)	110	82	85
Voltage ripple factor (full load) (%)	50∼100	<5	<1
Voltage ripple factor (current limit, interval) (%)	50∼200	2∼5	100∼200
Average operating electric field strength (MV/m)	0.15∼0.25	0.30∼0.40	0.25∼0.30
Adaptability to structure	Three-phase power source ≈ Single-phase power < High-frequency power source		
Adaptability to PM mass concentration	Three-phase power source < Single-phase power ≈ High-frequency power source		

three-phase power source is generally between 0.85 and 0.95, while the power factor of the single-phase power source is often low, about 0.6~0.8. In applications where single-phase power sources are converted to three-phase, the power factor can rise from 0.55 to 0.93. Therefore, the input power of the electrostatic precipitator can be greatly improved under the premise that the input power does not change significantly [17].

The high-voltage pulse power source realizes high-voltage (>120 kV) high-current (<300 A) discharge by voltage superposition principle, which can instantaneously excite high-concentration charge and force fine particles to charge, thereby effectively removing fly ash [19, 20]. The instantaneous power output of the pulse power source is relatively high. For example, in the research of Liu et al. [21], in a 300-MW coal-fired unit, the pulse width is only 75 μs. However, the excited charge concentration is several hundred times that of the conventional DC power source, which greatly increases the fly ash charge capacity, ensures the full charge of $PM_{2.5}$, and greatly improves the removal efficiency of $PM_{2.5}$. On the other hand, the narrow pulse makes the corona current to be released in a short time, which reduces the occurrence of back corona and further improves the PM removal efficiency. Third, the intermittent power source of the pulsed power source improves the efficiency of power utilization and has significant energy saving effects.

Zheng Qinzhen carried out a comparison of the control effects of different power sources in ESP on PM_{10} and $PM_{2.5}$ [22]:

Case A. Single-phase power source and three-phase power source

Two dual-chamber four-field electric precipitators were installed in a 660-MW power plant. The 16 single-phase power sources were upgraded into 16 three-phase high-efficiency power sources, and the total design capacity was reduced from 3376 to 2762 kW. During the test, the conversion factor of single-phase power source was 61%, and the conversion factor of the three-phase power source was 94%. After the three-phase power source was installed, PM_{10} and $PM_{2.5}$ at the ESP exit were reduced by 80.47% and 90.67%, respectively. The mass concentration of PM_{10} and $PM_{2.5}$ decreased from 62.43 and 23.83 mg/m^3 to 12.19 and 2.22 mg/m^3, respectively. The mass concentration ratio of $PM_{2.5}$ to PM_{10} at the outlet of the electrostatic precipitator decreased to 18.23% after being upgraded to the three-phase power source. The three-phase power source has higher collection efficiency for particles range from 0.2 to 0.5 μm.

Case B. High-frequency power source and single-phase power source

The experiment was carried out in two 660-MW power plants. Two double-chamber four-field electric precipitators were installed in these power plants. The parameters of structure, size, and rapping are basically the same, and the imported particle concentration is 20 g/m^3. One power plant is equipped with 16 high-frequency power sources, and the other is 16 single-phase power sources (experimental data in Case A). As for the PM_{10} and $PM_{2.5}$, the corresponding concentrations of the high-frequency power source were 59.85 and 6.56 mg/m^3, respectively. Due to the factors such as flue gas condition, boiler combustion, coal quality difference, operating parameters (high-frequency power conversion factor of only 53%), etc., the High-frequency power source has no significant change in the removal effect of large particles. However, the mass concentration ratio of $PM_{2.5}/PM_{10}$ decreased from 38.17% to 10.77%, indicating that the high-frequency power source has better control effect on fine particles.

Case C. Pulse power source and single-phase power source

600-MW unit, two double-chamber five-field PM collectors were used. The high-voltage power source uses 12 pulse power sources (superimposed 12 single-phase power sources) and 8 high-frequency power sources, with a power conversion factor of 71%. As for the concentrations of PM_{10} and $PM_{2.5}$, the corresponding concentrations of the pulse power source were 51.83 and 15.11 mg/m^3, respectively. The result is similar to the high-frequency power source. The mass concentration ratio of $PM_{2.5}/PM_{10}$ also decreased.

Liu et al. carried out the upgrading of the electric PM removal pulse power source in the 330-MW coal-fired unit [21]. The original electrostatic precipitator is a double-chamber four-field electric precipitator, and the power source is a high-frequency power source. The third and fourth electric fields of the electrostatic precipitator were upgraded as the pulse power source. Table 5.2 shows the comparison of particle and $PM_{2.5}$ emissions from the ESP outlet before and after the upgrading. It can be seen that the concentration of particles after the upgrading was 29.4 mg/m^3, and the concentration of $PM_{2.5}$ was 3.01 mg/m^3. The concentration of PM and $PM_{2.5}$ was 56.4% and 65.7% lower than that of the preupgraded, respectively. They also implemented a pulse power source upgraded for the double-chamber four-field electrostatic precipitator of the 660-MW

Table 5.2 ESP outlet emission before and after pulse power upgrading in 330-MW unit

Power source	PM (mg/m^3)	$PM_{2.5}$ (mg/m^3)
Pulse	29.4	3.01
High frequency	49.1	8.77

coal-fired unit, which was a power frequency power source before the upgrading. The emission of the electrostatic precipitator dropped from 70 to 16 mg/m^3, while achieving energy savings of 81%.

5.2.3 ESP structure upgrading

Optimizing the structure of the electrostatic precipitator can maximize the ESP index under the same size and improve its removal of $PM_{2.5}$. In order to improve the discharge characteristics of the electrostatic precipitator, the main structure upgrading of the energy-saving and emission reduction of the electrostatic precipitator is mainly focused on the first and second electric fields. For each additional partition, the electric field voltage can be increased by about 5–10 kV, and the electrostatic precipitator index can be increased by 10%–20%. In addition, the small partitions have the following advantages:(1) Improve the uniformity of the electrostatic PM discharge, reduce the back corona strength, and improve the effective PM collection volume of the electrostatic precipitator; (2) Reduce the re-entrainment of fly ash caused by the rapping.

Zheng Qinzhen carried out research on the influence of PM collection area on $PM_{2.5}$ for the five electric field electrostatic precipitator [22]. The PM collection area was 20% lower after turning off one of the electric fields. When any electric field was turned off, the PM concentration at the outlet of the electrostatic precipitator increased, and the emission concentration of the $PM_{2.5}$ at the electrostatic precipitator outlet increased. At the same time, the first and second electric fields and the final electric field had the greatest impact on $PM_{2.5}$ emissions.

Wang et al. carried out small-area upgrading and power source upgrading of the four electric field electrostatic precipitators [23]. The coal-fired power plant unit can achieve ultra-low emission. The concentration of PM_{10} and $PM_{2.5}$ after ESP was less than 15 and 2 mg/m^3, respectively.

5.2.4 Low-temperature economizer

The electrostatic precipitator has experienced three stages: high temperature (170~400°C, hot-side ESP), low temperature (130~170°C, cold-side ESP), and low-low temperature (70~130°C, colder-side ESP) [17]. Low-low temperature electrostatic precipitator is realized by setting a low-temperature economizer between the electrostatic precipitator and the air preheater. The low-temperature economizer is a gas-liquid heat exchanger that uses the heat of the flue gas from the flue to heat the boiler feed water, and is an effective means for energy saving and improving the efficiency of the electrostatic precipitator.

Under the same size of electrostatic precipitator structure, the influence of the low temperature of the flue gas has the following two aspects: On the one hand, the flow rate of the flue gas decreases as the temperature decreases, thereby increasing the specific PM collection area of the electrostatic precipitator; On the other hand, the specific resistance of the PM reduces as the temperature decreases, thereby the flashover frequency of the electrostatic precipitator reduces and the electrostatic precipitator index is improved. Deriving formula (5.2) shows that:

$$E_a \bullet E_p \bullet S = \psi \bullet \left(\frac{273 + T_0}{273 + T_1} \right)^3, \tag{5.3}$$

where T_0 and T_1 are the flue gas temperatures before and after cooling, °C; ψ is the working condition coefficient corresponding to the structure. The cooling can not only reduce the flow rate of the flue gas in the electrostatic precipitator, but also increase the gas density therein, thereby increasing the breakdown electric field strength and maximizing the electric PM removal index. Therefore, it can effectively improve the control effect of ESP on $PM_{2.5}$.

The PM_{10} and $PM_{2.5}$ emission in 330-MW coal-fired units installed with low-temperature economizers under different inlet temperatures had been compared [22]. The results are shown in Table 5.3, and the test temperatures were 160°C, 140°C, and 110°C, respectively. As the temperature decreased, the mass concentration of PM_{10} decreased. When the temperature was lowered from 160°C to 110°C, PM_{10} decreased by about 68%. The mass concentration of $PM_{2.5}$ decreased when the temperature was lowered from 160°C to 140°C, but from 140°C to 110°C, the concentration of $PM_{2.5}$ did not change significantly. According to the particle size distribution, the main reason was that the temperature of the particles from 120°C to 110°C, 0.2–1 μm particles significantly increased. However, with the decrease of

Table 5.3 ESP outlet emission under different inlet temperature conditions in 330-MW unit

Temperature	PM_{10} (mg/m^3)	$PM_{2.5}$ (mg/m^3)
160°C	47.90	7.16
140°C	23.75	3.59
110°C	15.50	3.86

temperature, the control effect of ESP on particles with aerodynamic diameter greater than 1 μm is significantly enhanced.

In summary, through the structure upgrading, the power source upgrading and the installation of the low-temperature economizer, the performance of ESP removes $PM_{2.5}$ can be promoted.

5.3 Impact of ESP-FF technology on $PM_{2.5}$

5.3.1 Mechanism of ESP-FF impact on $PM_{2.5}$

Electrostatic fabric filter composite PM removal technology (ESP-FF) is a new type of composite PM removal technology based on two kinds of mature PM removal theory: electrostatic PM removal and fabric filter PM removal [2]. It combines the advantages of electrostatic precipitator and fabric filter to improve the removal efficiency of $PM_{2.5}$, increase operating reliability, and reduce the PM removal cost of power plants. The fabric filter is a PM-removing device that mainly uses fabric filter elements to collect particles in flue gas. The main mechanism is that the PM is captured by the effects of sieving, inertia, adhesion, diffusion, and static electricity generated. The PM is captured by the action of screening or inertial force as it passes or bypasses the fabric filter. When the size of particles is lower than 0.2 μm, Brownian motion such as thermal motion of gas molecules is generated due to extremely fine particle. Since the fiber gap is smaller than the free path of the Brownian motion of the gas molecules, the particles are easily separated by the collision with the fibers. Collisions between particles can generate static electricity, which will charge the insulated filter cloth. In addition, the filtering action may be generated by the filter cloth itself or by particles accumulated on the filter cloth. Therefore, the ash accumulation on the filter bag will increase the capturing and diffusion effect of the particle collector, and the particles having a diameter much smaller than the filter material pore size can also be captured.

The electrostatic precipitator is suitable for collecting large particulate matter, and the fabric filter is suitable for collecting fine particles. In the ESP-FF, most of the fly ash in the flue gas is collected in the electric field, and then the remaining fine particulate matter is collected by the fabric filter. Electrostatic precipitator collects 80%~90% of the total particulate matter, and the fabric filter collects 10%~20% of the remaining particulate matter. The efficiency of the ESP-FF composite precipitator is not affected by the particle specific resistance and particle size, and is stable for a long time.

The predusting reduces the load of the fabric filter, thereby reducing the rate of resistance increase. Since the amount of particle collected in the fabric filter is small, the resistance rises slowly, so the fabric filter can be used longer than before, the energy consumption for cleaning particle can be saved, and the service life of the filter bag can be prolonged. Under normal conditions, the effect of the ESP-FF technology on the removal of $PM_{2.5}$ can reach 98%–99% [24, 25]. The technology is mainly divided into "pre-ESP post-FF" technology and embedded ESP-FF composite technology named advanced hybrid particulate collector (AHPC), which consists of a combination of fabric filtration and electrostatic precipitation in the same housing [26]. The latter one provided major synergism between the two collection methods, both in the particulate collection step and in the transfer of particle to the hopper. The AHPC provided ultrahigh collection efficiency, overcoming the problem of excessive fine-particle emission with conventional ESPs, and it solves the problem of re-entrainment and recollection of dust in conventional baghouses. Due to the electrocoagulation, the particle size increases. It is easier to be captured after entering the electric field, which overcomes the efficiency reduction brought by the cleaning process and improves the removal efficiency of $PM_{2.5}$.

The mechanism of highly efficient collection of $PM_{2.5}$ by ESP-FF composite precipitator can be described and summarized as shown in Fig. 5.3 [27].

The electric field area of ESP-FF charges the particles. The charge causes the particles to agglomerate and form larger particles. These larger particles are more easily trapped by electric fields. The charged particles that are not trapped by the electric field then flow to the fabric filter area

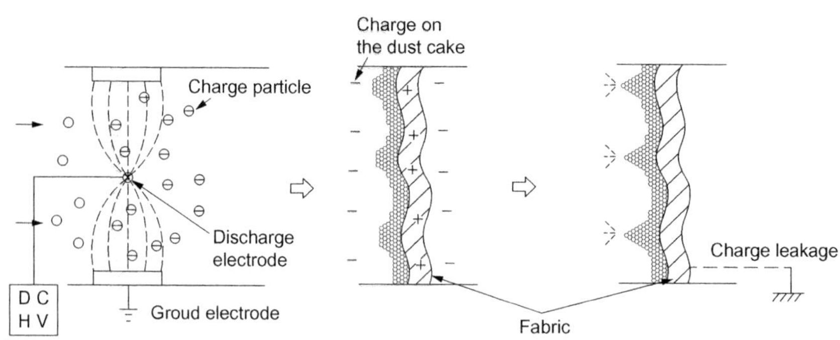

Fig. 5.3 Collecting mechanism of the ESP-FF.

re-electrostatically agglomerates before the filter bag. The agglomeration increases the capturing efficiency of the fabric filter on fine particles.

The PM layer adsorbed on the surface of the fabric filter is composed of the negatively charged PM. The PM with the same charge repels each other, making the structure of the PM layer loose, reducing the resistance of the PM layer, and reducing the penetration of fine particles. At the same time, these negatively charged PM layers form an electric field, which produces a certain electric field strength. These electric fields cause the PM to be polarized and electrophoresed to form a PM chain, thereby improving the collection efficiency of the fine particles.

5.3.2 The application of ESP-FF

Sun et al. carried out field tests on ESP-FF [28]. Unit 1 and 2 boilers of Zhuhai Power Plant are designed and manufactured by Mitsubishi Heavy Industries (MHI). The MB-FRR type boiler is a suspended 700-MW coal-fired boiler with 2290 t/h subcritical, intermediate reheat, forced circulation, and single furnace. The electrostatic precipitator is synchronously put into operation. In order to meet the ultra-low emission standard, the electrostatic precipitator was upgraded into an ESP-FF. The field test equipment for PM is DLPI. After the upgrading, the mass concentration of the emission from the precipitator can be maintained at 2.4–10 mg/m^3. Under the condition of full load condition and normal cleaning cycle, the removal efficiency of $PM_{2.5}$ in Units 1 and 2 is 99.12%, 99.44%, and the removal efficiency for PM_{10} is 99.38% and 99.62%, respectively. Fig. 5.4 shows the emission data of two units.

Chen et al. compared the $PM_{2.5}$ control effects of two power plants equipped with ESP-FF [29]. The boiler equipment in the A power plant is a 220 t/h circulating fluidized bed boiler, and the B power plant is a 1025 t/h circulating fluidized bed boiler, and both are denitrified by in-furnace calcium desulfurization and selective noncatalytic reduction (SNCR). The filed test used a DPI fine particle sampler manufactured by Dekati, Finland. Comparing the particle concentration data of the inlet, it was found that the removal efficiency of PM_{10}, $PM_{2.5}$, and PM_1 of A power plant reached 99.81%, 99.72%, and 99.37%, respectively. The B power plant also achieved similar removal efficiency, which were 99.81%, 99.63%, and 99.94%, respectively. In these power plants, most of the particulate matter was captured in the electric field, and the particles in the fabric filter area was substantially small size particles, so that the cake structure of the fabric filter

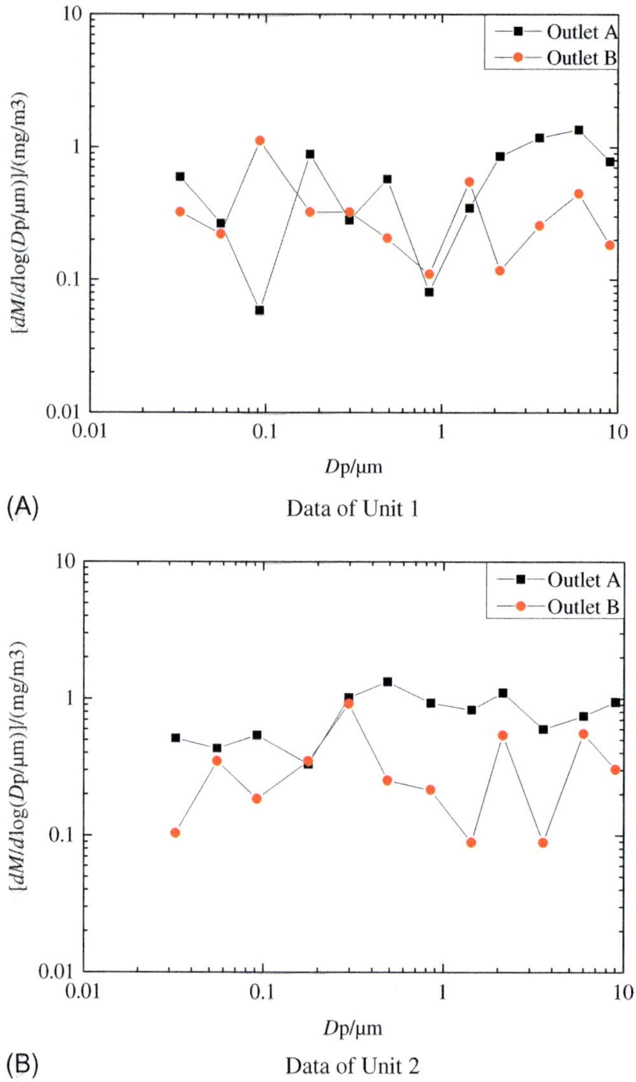

(A) Data of Unit 1

(B) Data of Unit 2

Fig. 5.4 Emission characteristics after application of ESP-FF technology.

area has a higher specific surface area. At the same time, due to the charged PM, the cake structure was fluffier than the fabric filter, which reduced the resistance of the particle collector.

Zhao et al. tested $PM_{2.5}$ in a pulverized coal furnace and a CFB boiler equipped with a fabric filter [30]. The experimental power plants are 220-MW pulverized coal boilers and 30-MW CFB boilers, all equipped with tandem ESP-FF. The study found that the removal efficiency of $PM_{2.5}$ was higher than 99.6%. The removal effect on the particles range from 0.1 to 1 μm was relatively low.

5.4 Impact of WFGD technology upgrading on $PM_{2.5}$

5.4.1 Upgrading function and difficulty of WFGD removing $PM_{2.5}$

Wet Desulfurization Technology (WFGD) is the main process for flue gas desulfurization in coal-fired power plants in China, mainly including limestone-gypsum wet method and seawater desulfurization. At the end of 2014, more than 94.4% of the desulfurization equipment was WFGD. The wet desulfurization process not only removes the SO_2 in the flue gas but also has a certain synergistic effect on the PM removal. The main principle is that some PM particles in the gas are captured in countercurrent contact with the droplets. The capturing mechanism mainly includes inertial collision, interception, and Brownian diffusion. [31]. Results of Wei et al. in the particle capture efficiency test carried out on 39 wet desulfurization devices showed the comprehensive PM removal efficiency of the wet desulfurization system was about 50%. The particle capture efficiency of the desulfurization system with high inlet particle concentration was higher than that of the low inlet particle concentration [32].

The new electrostatic precipitator technology reduces the concentration of particulate matter at its outlet. However, when the particle concentration at the outlet of the ESP, that is, the inlet of WFGD, is too low, the proportion of $PM_{2.5}$ in particles is greatly increased. For submicron particles, the inertial collision and interception effect is not obvious, and the decrease of the inlet particle size makes the PM removal efficiency of the desulfurization system greatly reduced [33]. Wang et al. studied five different units equipped with the wet desulfurization system and found that the mass concentration of $PM_{2.5}$ increased after WFGD, and the proportion of number concentration increased more obviously, indicating that the removal efficiency of wet desulfurization on $PM_{2.5}$ is much lower than PM_{10} [34]. At the same time, the particulate matter after the wet desulfurization system contains not only fly ash from coal combustion and fracture, but also new particulate matter produced by the desulfurization process. That is, the circulating slurry sprayed in the absorption tower will form $CaO/CaCO_3$, $CaSO_3/CaSO_4$, and fine droplets in the flue [33]. The field sampling test of Pan Danping also found the increase of $PM_{2.5}$ in the particles at the outlet of the wet desulfurization system, and the composition of these particles was basically gypsum and limestone [35]. Wei et al. believed that for lower particle concentration at the inlet of WFGD, the effect of droplets on outlet PM emissions is more pronounced [32]. This should be the reason for the higher concentration of particulate matter in the desulfurization outlet than that of the inlet.

In order to meet the ultra-low emission standard, the wet desulfurization system has been upgraded to enhance the removal effect on particulate matter, especially $PM_{2.5}$.

5.4.2 Integrated absorber technology

The integrated absorber technology is to improve the SO_2 removal efficiency through added new efficiency desulfurization devices or measures in the absorber. For examples, a gas-liquid mass transfer–enhanced device such as a turbulent flow–improved device, a tray device, or a bubbling-type device is installed between the slurry pool at the bottom of the desulfurization absorber and the spray layer on the upper portion or between the spray layers. By these methods, a stable liquid-holding layer is formed to increase the gas-liquid-solid three-phase mass transfer rate when the flue gas passes through the liquid-holding layer. At the same time, the gas-liquid distribution is improved by adjusting the spray density and atomization effect, thereby the SO_2 removal efficiency was improved. This desulfurization efficiency-enhancing measure also has the auxiliary function of synergistically capturing particulate matter in the flue gas. Combined with the high-efficiency mist eliminator or high-efficiency PM eliminator installed inside and outside the desulfurization absorber, the combined removal efficiency of the integrated absorber system can reach more than 70%. The processes currently used in this technology include spin-coupled WFGD technology, boiling foam, swirling bubbling, double trays, and turbulent tube grids [36, 37]. The spin-coupled desulfurization process was widely used in ultra-low emission projects. For example, Shanxi Yungang Thermal Power Unit 3 adopted the spin-coupled integrated absorber as a secondary PM removal. At the same time, the spray layer was upgraded and the tube bundle dust removal and mist eliminator were installed, so the particulate matter and SO_2 have meet the ultra-low emission standard [38].

Li et al. compared PM removal effects of desulfurization of different coal-fired units [39]. The particle size measurement of fine particles was carried out using a Finnish production of impact sampler (DPI) for sampling. Different desulfurization methods have some effects on PM removal, and the removal efficiency of PM_{10} is higher than that of $PM_{2.5}$, as shown in Table 5.4.

Zheng compared the effects of different spray layers on the removal of $PM_{2.5}$ (Table 5.5) [22]. In these tests, the particulate matter concentration of WFGD outlet remained unchanged. Under the condition of only two layers of spraying, the concentration ratio of $PM_{2.5}$ to PM_{10} was 30.64% in the WFGD outlet, and the ratio increased by 17.76% than that in the WFGD inlet. The capture efficiency of PM_{10} by WFGD was 52.0%, and $PM_{2.5}$ emissions were increased by 14.2%. When the four layers of the desulfurization tower were sprayed, PM_{10} and $PM_{2.5}$ decreased significantly after WFGD. The concentration ratio of $PM_{2.5}$ to PM_{10} was 29.2% in the WFGD outlet, and the ratio increased by 16.31% compared to that in the WFGD inlet. The increase of the number of spray layers was beneficial to improve the collection efficiency of the PM_{10} by the wet desulfurization absorber, especially the collection efficiency of the fine particles.

Zheng Qinzhen also reviewed the effects of different mist eliminators [22], as shown in Table 5.6. It can be found that the tubular mist eliminator and the three-stage roof demister have a great function of removing PM10 and $PM_{2.5}$.

Table 5.4 Particle removal efficiency under different desulfurization

Capacity (MW)	Desulfurization	Concentration at WFGD outlet (mg/m^3)		Removal efficiency of WESP (%)	
		PM_{10}	$PM_{2.5}$	PM_{10}	$PM_{2.5}$
300	Spin-coupled absorber	1.99	1.55	77.5	73.8
330	Double tray	5.31	4.59	47.4	36.1
600	Double tower double cycle	3.75	3.22	48.6	31.0
1000	Single tower double cycle	5.70	4.71	48.1	44.3

Table 5.5 Particle removal efficiency with different layers

Layers	Concentration at WFGD outlet (mg/m^3)		Removal efficiency of WFGD (%)	
	PM_{10}	$PM_{2.5}$	PM_{10}	$PM_{2.5}$
Two layers	6.87	2.11	52.0	Increased emission 14.2
Four layers	4.06	1.18	71.7	35.8

Table 5.6 Particle removal efficiency with different mist eliminators

Mist eliminators	Number	Concentration at WFGD inlet (mg/m^3)	Removal efficiency of WFGD (%)	
			PM_{10}	$PM_{2.5}$
Tubular	1	10~15	63.6	42.1
Mechanical grille	1	15	Increased	Increased
Two-stage roof demister	1	<8	Increased	Increased
Three-stage roof demister	3	15~20	75~80	60~65
		10~15	68.4	61.5

5.4.3 pH partition technology

There are two methods to achieve the pH partition, one is to set up two spray absorbers or install a separator in one spray absorber to physically partition the desulfurization slurry, the other is to form a natural partition depending on the characteristics of the slurry itself (flow direction, density, etc.). In this technology, the pH of one portion of the desulfurization slurry is maintained at a lower interval (4.5 to 5.3) to ensure limestone dissolution and the quality desulfurization gypsum, the pH of the other part of the desulfurization slurry is maintained at a higher range (5.8 to 6.4), which improves the absorption efficiency of SO_2 in the flue gas. At the same time, the desulfurization slurry spray (spray density, droplet size, etc.) can be optimized; on the one hand, it can improve the efficiency of desulfurization, on the other hand, synergistic capture of $PM_{2.5}$ in flue gas is also enhanced. Combined with a high-efficiency demister or PM demister installed inside and outside the desulfurization absorber, the synergistic removal efficiency of particulate matter in the pH partition system can reach 50% to 70%. Typical processes include single-absorber double pH, double-absorber double pH, single-absorber double partition, etc. [37].

Chen et al. conducted a sampling in a 600-MW ultra-low emission power plant, which installed a double spray absorber desulfurization equipment. The total particulate matter (TSP) in the flue gas was tested with Qingdao Laoying 3012H, and $PM_{2.5}$ was tested with DGI. After WFGD, the mass concentration of TSP decreased from 18 to 4.2 mg/m^3, and the removal efficiency was 76.67%, achieving ultra-low emission of particulate matter from coal-fired units. The mass concentration of $PM_{2.5}$ increased from 1.87 to 2.71 mg/m^3, which increased by 0.84 mg/m^3 after desulfurization [40].

It can be found that the WFGD system not only has no capturing effect on particles with a size of less than 1.0 μm, but also increases its mass concentration. The particles with a size of 0.5–1.0 μm increased by 0.035 mg/m^3, the particles with a size of 0.2–0.5 μm increased by 0.33 mg/m^3, and the particles with a size of less than 0.2 μm increased by 1.26 mg/m^3.

5.5 Impact of WESP on $PM_{2.5}$

5.5.1 Mechanism of wet electrostatic precipitator on $PM_{2.5}$

Wet electrostatic precipitator (WESP) has been used in coal-fired power plants for nearly 30 years, and its technology is relatively mature. It is estimated that the number of different types of WESPs used in coal-fired power plants in the United States, Japan, and Europe has exceeded 100. In order to reduce the particulate matter produced by coal combustion and the secondary particulate matter produced by the desulfurization process, WESP has become the terminal precision processing equipment for coal-fired power flue gas pollution control in China in recent years. It mainly controls $PM_{2.5}$, SO_3 acid mist, and gypsum rain, thus achieving ultra-low emission of particle.

The principle of the wet electrostatic precipitator is introduced as follows: the metal discharge line ionizes the surrounding gas under the action of DC high voltage, so that the surface of the PM or droplet particles is charged. The charged particles move to the PM collecting plant because of the electric field force and then deposit on the PM collecting plate. The water flows down from the top of the PM-collecting plate, forming a uniform and stable water film on the PM-collecting plate, and taking away the PM on the plate. Therefore, the wet electrostatic precipitator has the same principle as the dry ESP, and both of them have the same three stages, which are charging, collecting, and cleaning.

However, unlike ESP cleaning, the wet electrostatic precipitator uses liquid to flush the surface of the PM collecting plate, thereby the problem of re-entrainment caused by the rapping is avoided. At the same time, WESP can provide several times the corona power of that dry PM collector, which can improve the collection efficiency for $PM_{2.5}$. Depending on the electrode, WESP can be divided into metal plates, conductive glass plates, and flexible plates [41].

WESPs have a removal effect on SO_3 which is the precursor of $PM_{2.5}$. On the one hand, the SO_3 acid mist will coagulate and grow in the electric field, and then will be captured on the collecting plate by the liquid film. On the other hand, the SO_3 acid mist is deposited on the surface of the fine particles to effectively reduce the specific resistance of the particles, improving the PM removal efficiency and thereby also improving the SO_3 removal efficiency.

There is a stable removal efficiency on PM for the wet electrostatic precipitator. If it is difficult to achieve the ultra-low emission of PM, WESP can be combined with the dry ESP and WFGD, or be combined with low-temperature ESP, or ESP-FF, and so on to capture PM.

5.5.2 Application of wet electrostatic precipitator

Li et al. focused on 13 sets of ultra-low-emission units with different types of wet electrostatic precipitators to test their effects on particulate matter emissions [42]. The results showed that the TSP concentration at WFGD outlet can reduce to less than 5 mg/m^3 and soluble salt included in the droplets can reduce to less than 2.6 mg/m^3 in the ultra-low emission coal-fired units. If WESP were to be added, the removal efficiency of $PM_{2.5}$ can reach 70% or above, and soluble salt included in the droplets can reduce to less than 1.4 mg/m^3. The detail is shown in Table 5.7.

As shown in Table 5.7, the TSP concentration at the outlet of the wet electrostatic precipitator ranges from 2.1 to 4.9 mg/m^3, and the removal efficiency of TSP was over 80%. The mass concentration of $PM_{2.5}$ ranges from 1.1 to 3.6 mg/m^3, and the removal efficiency of $PM_{2.5}$ was above 78%. The concentration of SO_3 in the WESP inlet was 14~55 mg/m^3, the concentration of SO_3 in the outlet was 2.5~10 mg/m^3, and the removal efficiency of SO3 was 65%.

Liu et al. compared the $PM_{2.5}$ emissions of 15 ultra-low emission power plants equipped with wet electrostatic precipitators. The results are shown in Figs. 5.5 and 5.6 [43].

Most of the $PM_{2.5}$ data tested by Liu et al. were less than 1 mg/m^3, and the WESP removal efficiency was mostly above 60%. The relationship between the TSP and $PM_{2.5}$ mass concentration is shown in Fig. 5.6. The lower the total PM concentration at the exit, the higher the proportion of $PM_{2.5}$, and the highest datum is 75%.

Zhengda et al developed a new WESP system with the assistance of a precharger [44]. The technology was used in a 1000-MW coal-fired power

Table 5.7 The removal efficiency of WESP on TSP, $PM_{2.5}$, and SO_3

		TSP			$PM_{2.5}$			SO_3		
Item	Capacity (MW)	Inlet (mg/m^3)	Outlet (mg/m^3)	Removal efficiency (%)	Inlet (mg/m^3)	Outlet (mg/m^3)	Removal efficiency (%)	Inlet (mg/m^3)	Outlet (mg/m^3)	Removal efficiency (%)
Unit 1	300	14.5	2.6	82.07	10.5	1.2	88.93	16.0	4.2	73.75
Unit 2	300	13.2	2.1	84.09	9.5	1.1	88.97	17.3	4.7	72.83
Unit 3	330	47.4	4.6	90.28	30.7	2.6	91.63	20.8	4.1	80.16
Unit 4	330	49.6	4.8	90.40	23.1	3.4	85.50	30.6	2.5	91.84
Unit 5	330	36.3	4.4	87.92	24.5	2.2	91.12	14.8	3.2	78.58
Unit 7	330	37.4	4.2	88.77	16.9	3.6	78.70	19.1	4.4	76.96
Unit 8	330	16.4	3.0	81.71	9.3	1.6	83.28	20.4	2.0	90.20
Unit 9	330	44.9	4.9	89.09	14.5	1.9	88.16	46.8	4.6	90.17
Unit 10	600	28.9	2.9	89.90	13.9	1.6	88.16	26.6	8.7	67.41
Unit 11	660	38.6	4.6	87.99	13.9	1.6	88.16	34.7	8.2	76.28
Unit 12	660	24.6	2.6	89.43	22.3	2.0	91.26	51.5	7.2	86.02
Unit 13	670	41.1	4.0	90.27	18.6	2.7	85.78	26.7	5.0	81.36

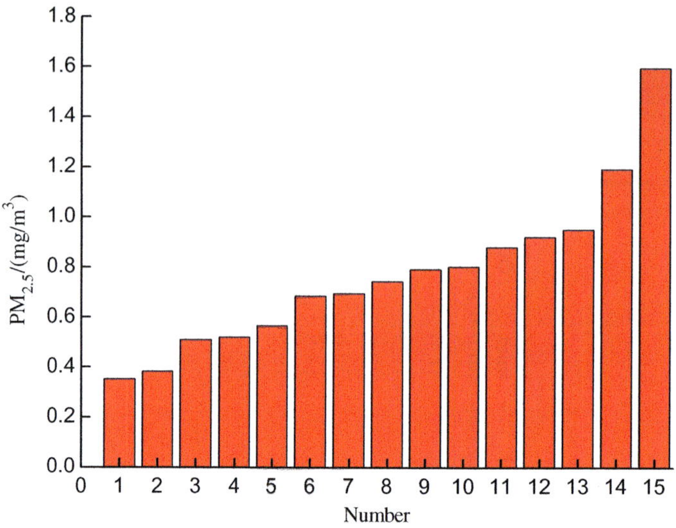

Fig. 5.5 Mass concentration of $PM_{2.5}$ in the outlet of WESP.

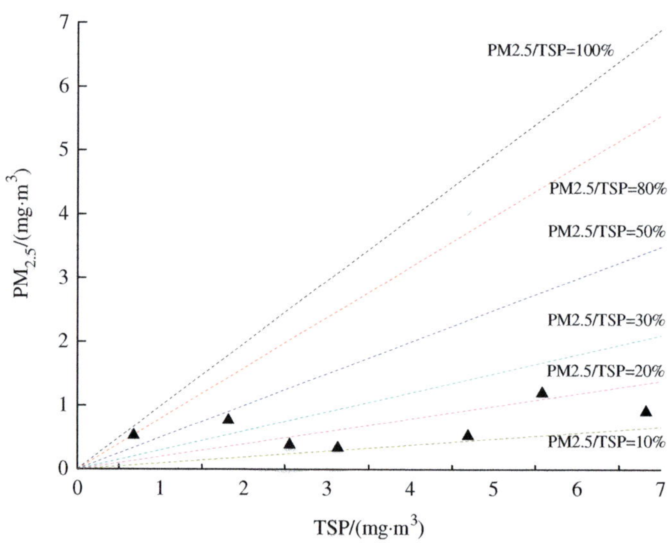

Fig. 5.6 The relationship of TSP and $PM_{2.5}$ in the WESP outlet.

plant, the average PM emission level was 1.18 mg/m^3 before upgrading with the precharger. The rapping of dry ESP may lead to a rising concentration at the WESP inlet, and the corresponding emission concentration could rise to ~3 mg/m$^{3.}$ After installing the precharger, the average PM emission level was reduced to 0.43 mg/m^3.

In summary, although WESP removal efficiency of $PM_{2.5}$ does not reach 99.9% like electrostatic precipitator, it is the last gatekeeper of Air pollution control devices (APCDs), and it is the key device to achieving ultra-low emissions of PM for coal-fired power plants.

5.6 Case of $PM_{2.5}$ changes in an ultra-low emission power plant

Sampling experiments were conducted in Sanhe (SH) Power Plant Unit 4. The unit had a capacity of 300 MW. The proximate and ultimate analysis of the coal is presented in Table 5.8. The unit had been upgraded to meet the ULE standards, based on the technology route shown in Fig. 5.7. The particulate emission control devices (PECD) installed at the SH unit are listed in Table 5.9. After the LLT ESP system was installed, the PM concentration decreased firstly, then improvements to the FGD decreased the PM concentration secondly, and the installation of the WESP decreased PM emissions finally. The PM measurement for each control device was simultaneous, so the ESP, FGD, and WESP data can be compared.

The PM samples were collected and analyzed by an ELPI+ (Dekati Inc.), which focused on PM_{10} particles. This collection and analysis method have been widely used in lab scale and full-scale experiments. Bulk PMs are removed by the cyclone cutter through a heated sampling probe as specified in EPA standard 201 A. The dilution ratio for the dilutor is set at 8. A Perma Pure gas dryer removes moisture that might affect the measurement results from the sample gas. The probe, cyclone cutter, and the first diluter are heated to 120°C to prevent condensation. The sampling positions and the gas pretreatment system are shown in Table 5.10. A polycarbonate membrane was used to collect the PM samples that were analyzed using SEM/EDS (JEOL JEM-6490LV).

Table 5.8 Coal analysis of SH #4 unit (wt.%, dry basis)

Proximate analysis			Elemental analysis				
Volatile	Ash	Fixed carbon	C	H	N	S	O
29.04	18.58	52.38	63.22	4.18	0.89	0.74	12.39

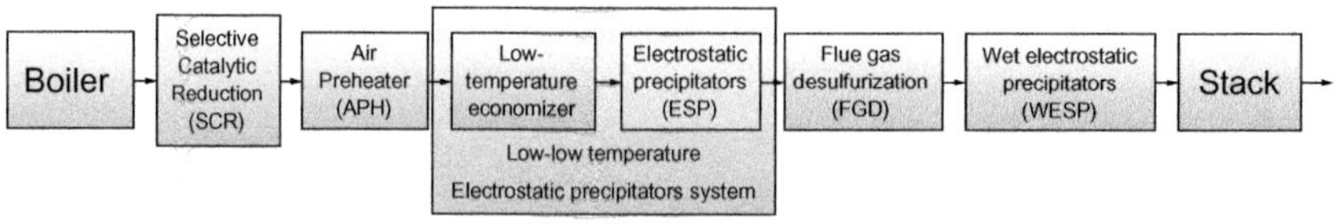

Fig. 5.7 Technology route for ULE upgrading.

Table 5.9 Air pollution control device upgrading of SH Unit 4

	LT-economizer	ESP	FGD	WESP
Action	New construction	LLT ESP, high-frequency electric source upgraded	Integrated desulfurization and dedusting upgraded	New construction

Table 5.10 Sampling locations and flue gas pretreatment

	Sampling position			
	ESP inlet	ESP outlet	FGD outlet	WESP outlet
PM concentration	High	Slightly low	Low	Low
Humidity	Low	Low	High	High
Flue gas pretreatment system	PM_{10} cyclone + two-stage dilutor	PM_{10} cyclone + one-stage dilutor	One-stage dilutor + Perma Pure gas dryer	One-stage dilutor + Perma Pure gas dryer

5.6.1 PM concentrations

The mass concentrations and numbers of particles in the SH unit flue gas are presented in Fig. 5.8. Prior to entering the ESP, the mass concentrations of PM_{10}, $PM_{2.5}$, and PM_1 (PM less than 1 μm at the aerodynamic diameter) were 1087.95 mg/m^3, 454.76 mg/m^3, 82.28 mg/m^3, respectively. After exiting the WESP, the mass concentrations of PM_{10}, $PM_{2.5}$, and PM_1 were 0.36 mg/m^3, 0.36 mg/m^3, and 0.09 mg/m^3, respectively (Fig. 5.8A).

The ratio of $PM_{2.5}$ mass-concentration to PM_{10} mass-concentration ($PM_{2.5}/PM_{10}$) is commonly used to assess the effects of different technology on PM control. After the ESP, the $PM_{2.5}/PM_{10}$ and PM_1/PM_{10} ratios decreased to 17.2% and 4.3%, respectively. These numbers indicate that the upgraded ESP can effectively control $PM_{2.5}$. However, the $PM_{2.5}/PM_{10}$ ratio and the PM_1/PM_{10} ratio increased to 62.6% and 19.4% after

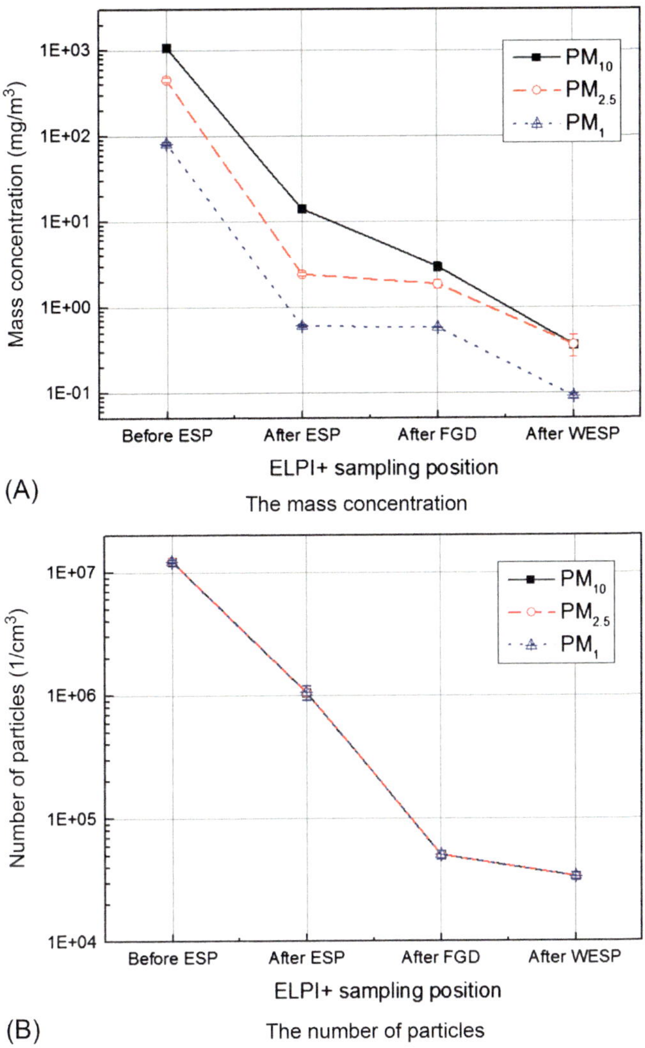

Fig. 5.8 Concentrations of fine PM in ESP, FGD, and WESP.

the FGD, respectively. After the WESP, the $PM_{2.5}/PM_{10}$ ratio is close to 1 and the PM_1/PM_{10} ratio increased to 24.9%. The results indicate that WESP will effectively remove PM between 2.5 μm and 10 μm in flue gas and that most of the PM emitted to the environment will be less than 2.5 μm in aerodynamic diameter (D_a).

As shown in Fig. 5.8B, the curve for the number of PM_{10}, $PM_{2.5}$, and PM_1 particles overlaps. This is because the number of particles between 1 μm and 10 μm is less than 1% of the number of PM_{10}. The number of

PM_{10} particles decreased from 1.24E+7 to 3.41E+4 per cubic centimeter. When the flue gas goes through the ESP and FGD, the number of particles was reduced by 91.4% and 95.2%, respectively, indicating that ESP and FGD can effectively control ultrafine PM.

In this case, the PM concentration before the ESP was set to 100%. Then, the PM distribution at each sampling point was calculated, as shown in Fig. 5.9. For the mass distribution, 0.033% of PM_{10}, 0.080% of $PM_{2.5}$, and 0.11% of PM_1 leave the power plant, respectively (Fig. 5.9A). For the number distribution of PM, 99.7% of PM was captured by the upgraded ESP, FGD, and WESP (Fig. 5.9B). Thus, the combined application of

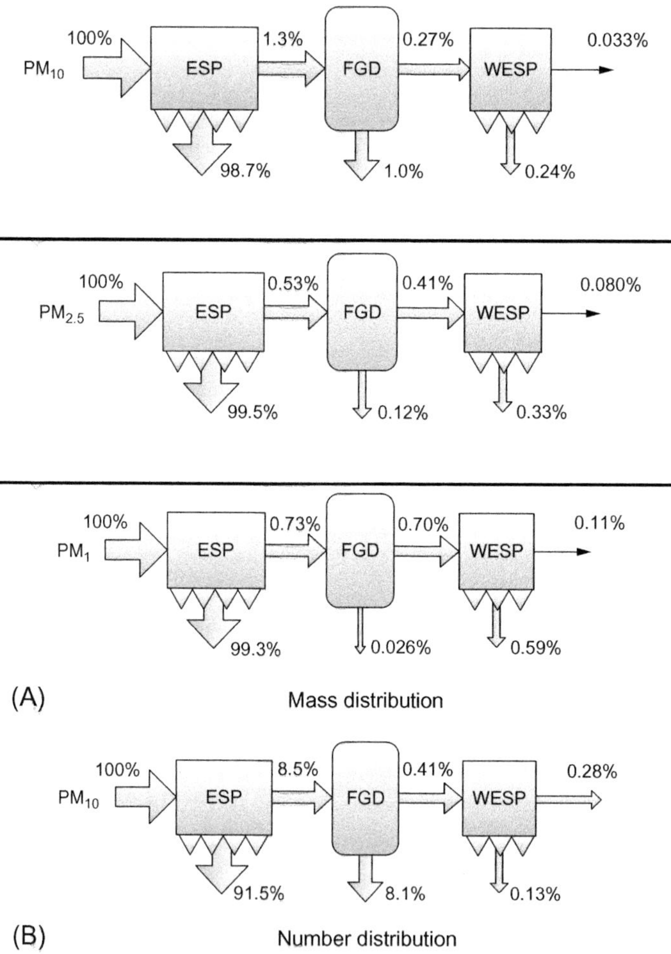

Fig. 5.9 Fine PM distribution in ESP, WFGD, and WESP.

low-temperature economizer, and upgraded ESP, FGD, and WESP significantly control the PM. In other words, over 99.7% of the particles with aerodynamic diameter less than 10 μm generated by coal combustion were captured by the PECD in the SH unit. For the mass distribution, more than 98.7% of fine particles were captured by the ESP. The mass distribution and the number distribution $PM_{2.5}$ collection efficiencies for traditional ESPs were usually less than 98.0% and 91.5%, respectively [45]. This indicates that the LLT ESP system is an improvement over the traditional ESP.

5.6.2 PM size distribution

ELPI+ can collect real-time concentration data at different D_a ranges. This D_a information was used to examine the size changes in PM_{10} at ESP, WFGD, and WESP locations.

Before the ESP, two peaks were observed at D_a equal to 0.177 and 5.98 μm in the mass-concentration data. One peak was observed at 0.0561 μm in the number of particles data (shown in Fig. 5.10). There was a peak at 0.0134 μm after the ESP in both the mass-concentration data and the number of particles data. The PM between 0.0134 and 0.0212 μm increased in the ESP. This effect has been mentioned in other research and could be from larger particles breaking down into smaller particles in the ESP system. PM less than 0.03 μm was poorly controlled by the ESP.

Wet limestone-gypsum desulfurization (WFGD) follows the ESP in the SH unit. One peak can be observed at 2.14 μm in the mass-concentration data and one peak at 0.561 μm in the number of particles data (shown in Fig. 5.11). The data show a significant decrease in PM less than 0.03 μm in the WFGD. It is likely that the scrubbing effect of WFGD assists in the control of smaller particles not removed by the ESP.

After the WESP, two peaks were observed at D_a equal to 0.848 μm and 2.14 μm in the mass-concentration data. While one peak was observed at D_a equal to 0.561 μm in the number of particles data (shown in Fig. 5.12). The data indicate that particles in a wide size range are removed by the WESP system.

5.6.3 PM capture efficiencies of the ESP, FGD, and WESP

The effectiveness of a PECD can be measured by calculating the capture efficiency, as shown in Eq. (5.4).

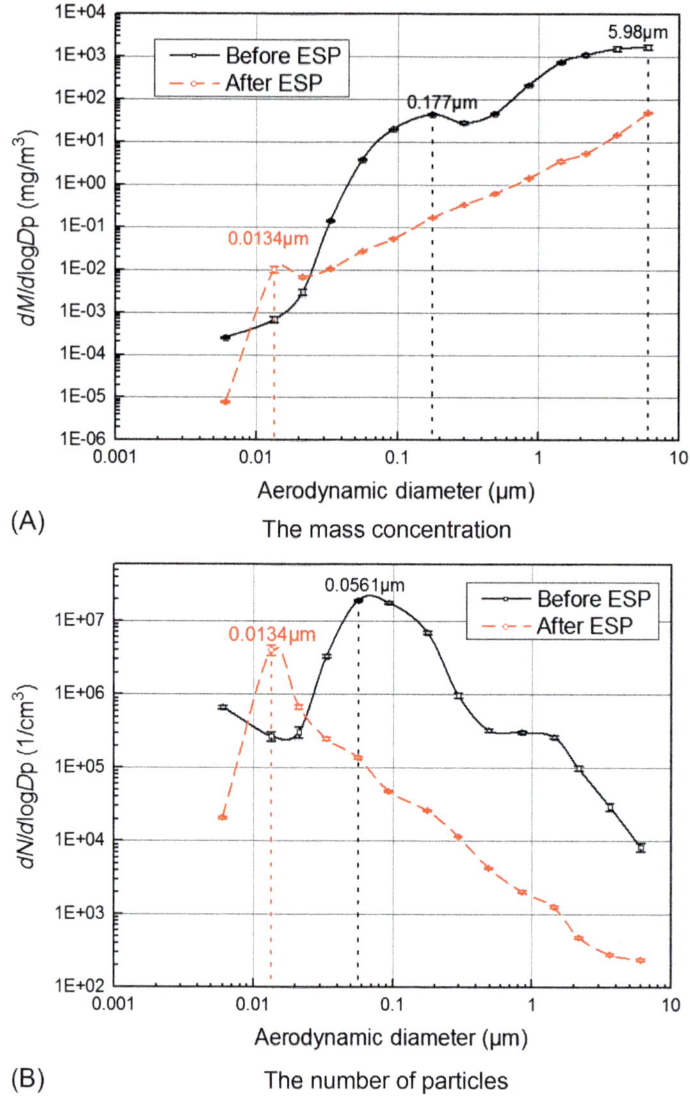

Fig. 5.10 PM size distribution before and after ESP.

$$\eta = \frac{C_{\text{in}} - C_{\text{out}}}{C_{\text{in}}} \times 100\%, \tag{5.4}$$

where η represents the capture efficiency, C_{in} represents the concentration of PM before the device, and C_{out} represents the concentration of PM after the device.

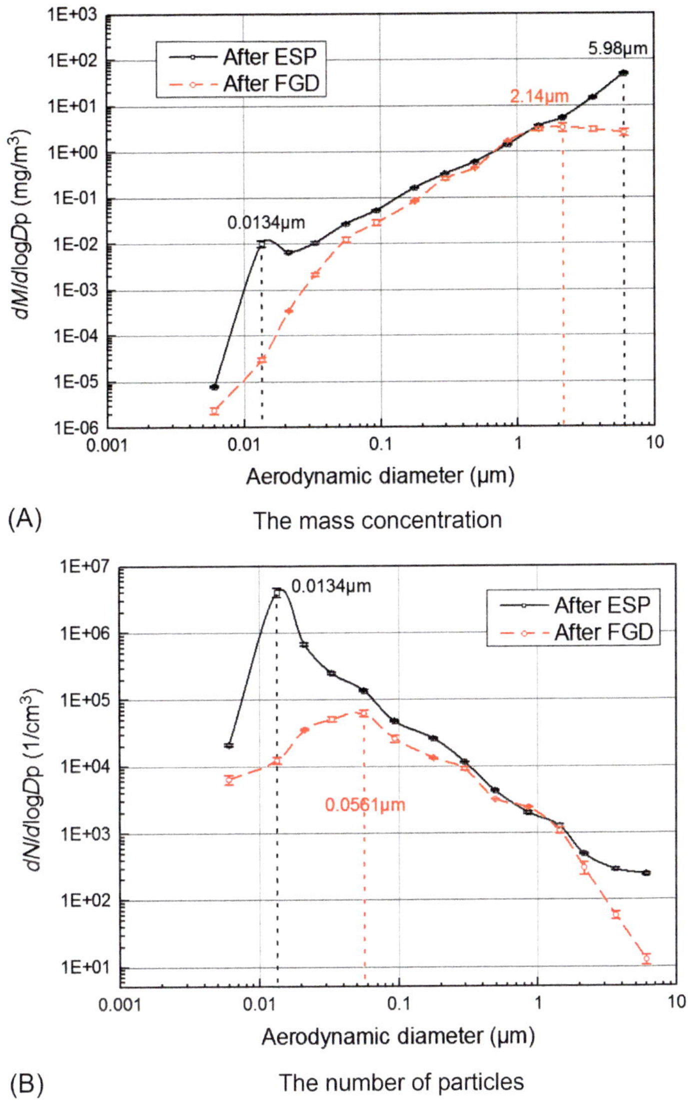

Fig. 5.11 PM size distribution before and after FGD.

The capture efficiencies of $PM_{10}/PM_{2.5}/PM_1$ in the ESP, WFGD, and WESP are shown in Fig. 5.13. The PM_{10} mass-concentration capture efficiencies (η_M) in the ESP, WFGD, and WESP were 98.7%, 78.9%, and 87.8%, respectively. The $PM_{2.5}$ and PM_1 mass-concentration capture efficiencies were high for the ESP and WESP and low for the WFGD. Altman

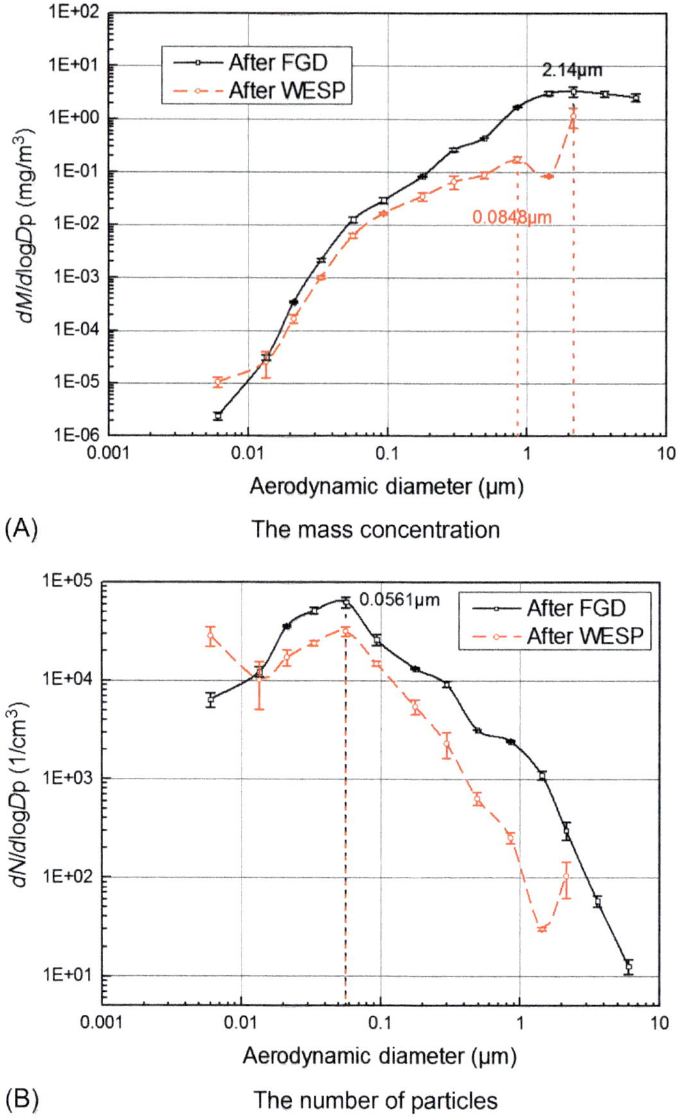

Fig. 5.12 PM size distribution before and after WESP.

et al. determined that the $PM_{2.5}$ capture efficiency of the WESP was more than 99.9% [46]. In this test, the WESP capture efficiency was only 80.5%. This may be due to the low concentration at the WESP inlet. Thus, the minimum mass concentration at the WESP outlet would be between 0.1 and 1 mg/m^3.

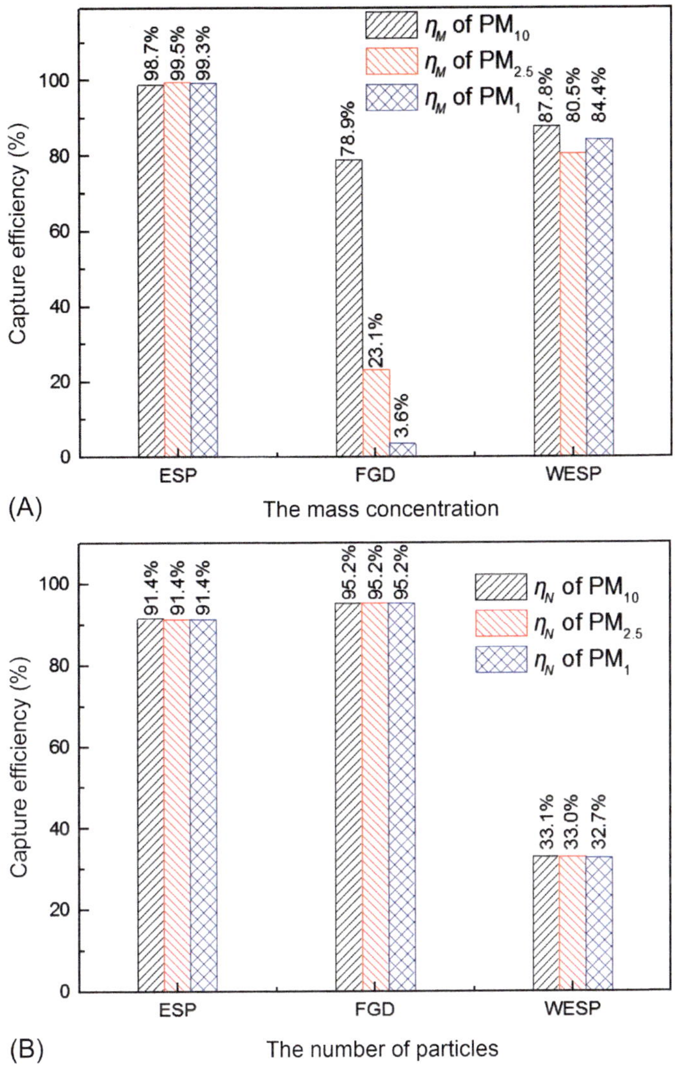

Fig. 5.13 Concentration and capture efficiencies of $PM_{10}/PM_{2.5}/PM_1$ in ESP, WFGD, and WESP.

Because the number of particles between 1 and 10 μm was less than 1% of the number of PM_{10} particles, the number capture efficiencies (η_N) of PM_{10}, $PM_{2.5}$, and PM_1 were all about the same. The η_N of PM for the ESP is about 91.4%. Therefore, the ESP effectively controls particles larger than 0.03 μm. The number capture efficiency at the WFGD location was calculated as 95%. The effective control at the WFGD is likely related to the scrubbing effect of the WFGD on the PM less than 0.03 μm. The number capture efficiency at the WESP location was only about 33%. The data indicate that ESP and WFGD are better at controlling ultrafine PM ($PM_{0.1}$) than WESP.

5.6.4 Capture efficiencies of PM at different D_a

Because the mass concentration and number of particles are all proportional to the measured current of each stage, the calculated mass-concentration capture efficiencies and number of particle capture efficiencies were the same. The relationship between capture efficiency (η) and aerodynamic diameter (D_a) is shown in Fig. 5.14. A negative capture efficiency indicates an increase in the PM in this size range. The ESP of the SH unit had excellent capture efficiency for PM larger than 0.03 μm (Fig. 5.14A). The capture efficiencies of PM larger than 0.05 μm and less than 5 μm in D_a were all close to 99%. The synergic effect of the LT-economizer and the upgraded ESP greatly increased the capture ability of PM larger than 0.05 μm. This also explains the decrease in the $PM_{2.5}/PM_{10}$ ratio. However, there was a negative capture efficiency for PM between 0.01 μm and 0.03 μm. This means that there are more small particles at the ESP outlet than at the ESP inlet. This is most likely due to the generation of smaller particles from the breakdown of larger particles and poor capturing capability for particle sizes less than 0.03 μm in the ESP system.

The η at the WFGD location, as shown in Fig. 5.14B, decreases initially and then increases with increasing D_a. The capture efficiencies of PM between 0.05 and 2.5 μm were very low. The PM near 1 μm in size increased only a little, while the η of PM at other D_a ranges was above 70%. This is consistent with the mass-concentration control effects observed at the WFGD. Although the mass-concentration of PM_{10} was greater than $PM_{2.5}$ and PM_1, the number of PM_{10} particles is effectively controlled (η_N is close to 95%). The WFGD system removes part of the coal-fired PM; however, the limestone-gypsum particles can be carried into the flue gas resulting in a net low capture efficiency of PM between 0.05 μm and 2.5 μm in D_a. The relationship between η and D_a in the WESP is shown in Fig. 5.14C. With decreasing D_a, η trended downward. PM greater than 0.3 μm were well controlled by the WESP. The η of $PM_{0.02-0.2}$ remained essentially stable at around 50% and decreased with D_a less than 0.02 μm. The increase of PM less than 0.01 μm may be due to ELPI+ operation conditions and the

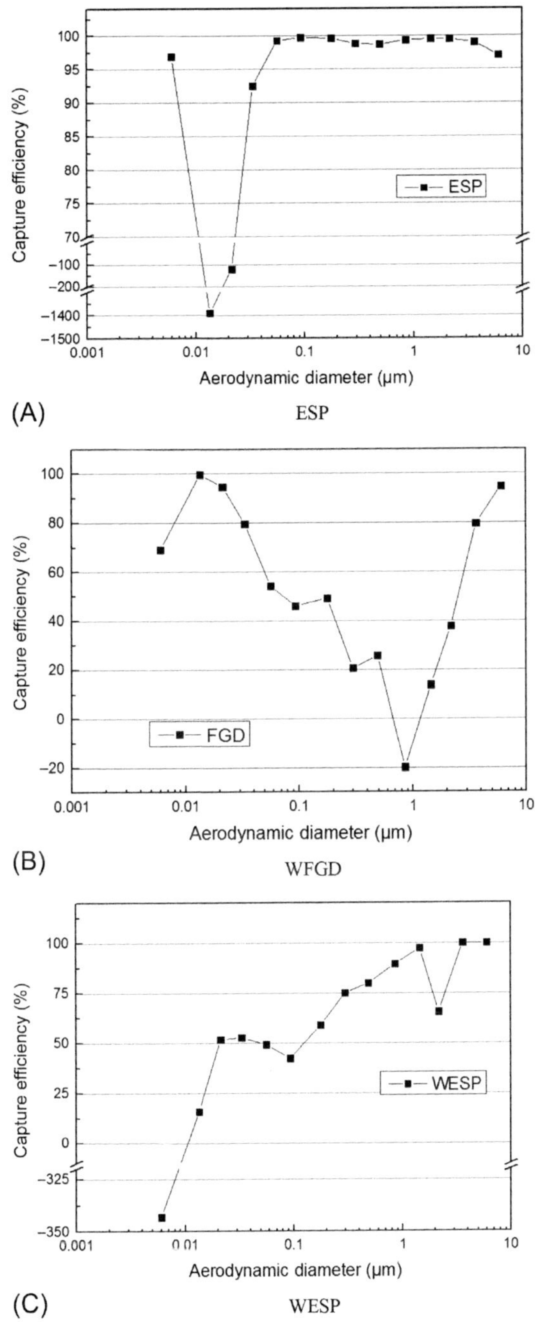

Fig. 5.14 The capture efficiencies of fine PMs at different D_a.

influence of moisture in the sampling gas. The WESP was the preferable final control device in a ULE power plant to further remove submicron PM and PM less than 0.3 μm.

5.6.5 SEM/EDS analysis

The PM samples were analyzed by SEM/EDS, and images are shown in Figs. 5.15 and 5.16. The PM samples before and after the ESP were mainly spherical. However, after the ESP, electric agglomeration of fine particles during the charging process caused ultrafine particles to attach to the particle surface (Fig. 5.15A and B). After the WFGD, the PM was present as either spherical or columnar particles. The spherical particles were composed of aluminosilicate material from coal burning and the columnar particles were gypsum particles (Fig. 5.15C). The CaO/$CaCO_3$ sulfate particles likely adhered together and were collected by ELPI+ (Fig. 5.15D). This same effect was observed in other literature [Wang H, Song Q, Yao Q, Chen CH. Experimental study on removal effect of wet flue gas desulfurization system on fine particles from a coal-fired power plant. Proc. Chin. Soc. Electrical Eng. 2008;28(5):1–7 (In Chinese)]. The columnar particles decreased after the WESP (Fig. 5.15E and F).

As shown in Fig. 5.16A, S mainly exists in ultrafine PM ($PM_{0.2}$). After the ESP, the S mass fraction of PM between 0.5 and 2.5 μm in D_a increased slightly due to electric agglomeration. The S mass fraction increased after the WFGD and decreased after the WESP for D_a between 0.3 and 1.5 μm. This was in agreement with the change in Ca mass fraction (Fig. 5.16B). The limestone/gypsum particles were carried into the flue gas after the WFGD and subsequently removed by the WESP. As shown in Fig. 5.16C–E, Na, K, and Mg were mainly present in PM with D_a between 0.2 and 2.5 μm. This is most likely due to their low condensation temperature and adhesion to the submicron particle surface. The Cl mass fraction significantly increased after the WFGD (Fig. 5.16F) because of the high Cl content in the circulating water. In summary, the PM emissions in the SH unit consist mainly of coal-burning particles mixed with a much smaller amount of limestone/gypsum droplets.

In summary, the addition of LLE and WESP, upgrading of ESP and WFGD, all these technologies measured will not only increase the capture efficiency on PM or SO_2, but also can reach have a positive effect on removing PM_{10} and $PM_{2.5}$.

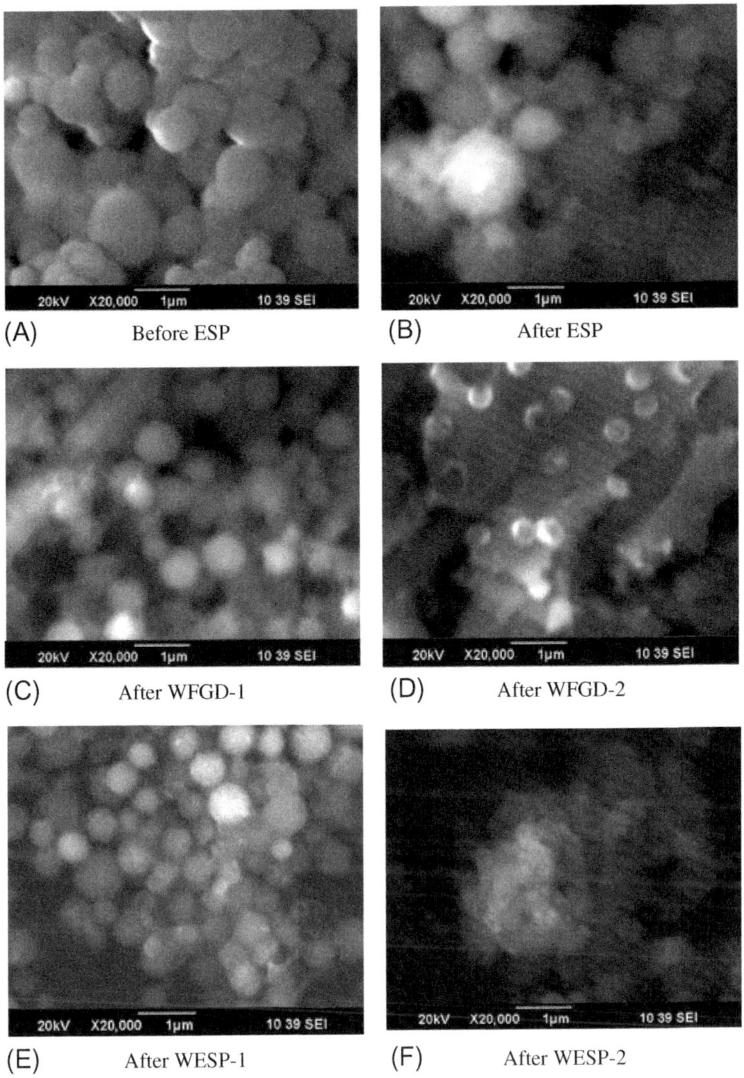

Fig. 5.15 SEM images of particles sampling by ELPI+.

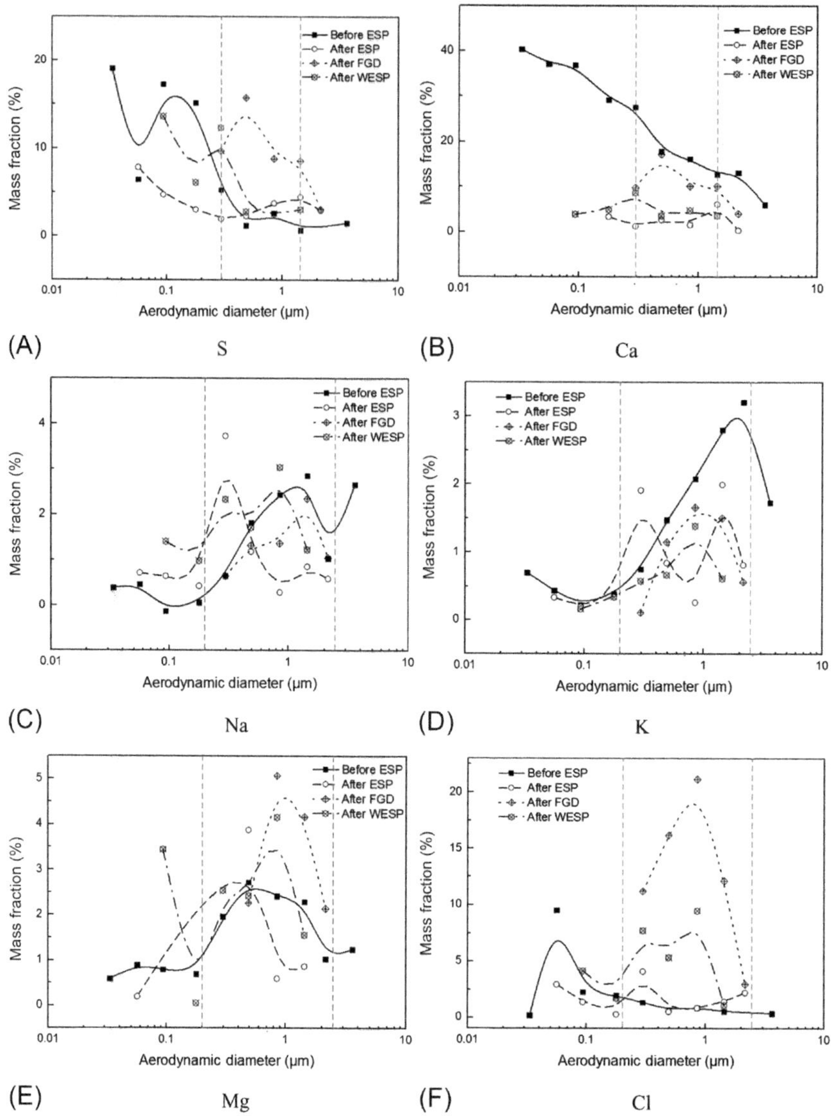

Fig. 5.16 Chemical compositions of each stage of samples derived from ELPI+ by the SEM/EDS.

References

[1] J.J. Cao, PM2.5 and the Environment in China, Science Press, China, 2014.

[2] L.J. Yang, Pollution Control Technology of Fine Particulate Matter from Combustion Source, Chemical Industry Press, China, 2011.

[3] D.X. Yu, M.H. Xu, F. Yi, J.H. Huang, G. Li, A review on particle formation mechanisms during coal combustion, Coal Convers. 27 (4) (2004) 7–12.

[4] IEA, Energy and Air Pollution: World Energy Outlook Special Report, 2016.
[5] Y.H. David, S.C. Chen, Z.L. Zuo, $PM_{2.5}$ in China: measurements, sources, visibility and health effects, and mitigation, Particuology 2 (3) (2014) 1–26.
[6] J.H. Yi, X.F. Wu, L.Y. Wang, J. Lei, J. Han, Mechanisms and effects of $PM_{2.5}$ on diseases of respiratory system: latest findings, J. Xi'an Jiaotong Univ. (Med. Sci.) 40 (1) (2019) 175–180.
[7] J.G. Li, Y. Liu, The effect of compositions of domestic coal on the performance of ESP and an evaluation of ESP adaptability, Sci. Technol. Rev. 28 (7) (2010) 104–109.
[8] H.J. White, Industrial Electrostatic Precipitation, Addison Wesley, Boston, 1963.
[9] J.F. Hughes, Applied Electrostatic Precipitation, Springer, Netherlands, 1998.
[10] G.J. Chen, Dust control technologies under the new emission standard for thermal power plants, Sci. Technol. Rev. 28 (3) (2010) 90–95.
[11] J.M. Hao, L. Duan, H.H. Yi, Physicochemical Characteristics of Inhalable Particulate Matter From Combustion Source, Science Press, 2008.
[12] S.L. Wang, Y. Chen, P. Han, G.Q. Zhang, Q.Z. Zheng, X.J. Shen, S.R. Li, K.P. Yan, PM_{10} and $PM_{2.5}$ emission control by electrostatic precipitator (ESP) for coal-fired power plants I: ESP sizing and applications, Sci. Technol. Rev. 32 (33) (2014) 23–33.
[13] A. Marquard, J. Meyer, G. Kasper, Characterization of unipolar electrical aerosol chargers—part I: a review of charger performance criteria, J. Aerosol Sci. 37 (9) (2006) 1052–1068.
[14] S. Oglesby, Electrostatic Precipitation, Marcel Dekker INC Press, New York, 1978.
[15] J. Zhu, Q. Zhao, Y. Yao, Effects of high-voltage power sources on fine particle collection efficiency with an industrial electrostatic precipitator, J. Electrost. 70 (3) (2012) 285–291.
[16] K.P. Yan, Electric precipitation upgrading of coal-fired power plant: basic principle and industrial application, in: Proceedings of the 17th Conference of Electrostatic Precipitation, Chinese Society of Electrostatic Precipitation, 2017.
[17] K.P. Yan, S.R. Li, Q.Z. Zheng, J.X. Zhou, Y.F. Huang, Z. Liu, Development and application of electrostatic precipitation technology, High Voltage Eng. 43 (2) (2017) 476–486.
[18] J.B. Zhu, Fine Particle Collection With Electrostatic Precipitation, Zhejiang University, 2010.
[19] G. Norbert, H. Werner, K. Michael, Application of different types of high-voltage supplies on industrial electrostatic precipitators, IEEE Trans. Ind. Appl. 40 (6) (2004) 1513–1520.
[20] V. Reyes, P. Elholm, 4th generation of Coromax pulse generators for ESPs, in: International Conference on Electrostatic Precipitation (ICESP), 2011, pp. 1–7.
[21] Y.F. Liu, A new type of high voltage pulse electrostatic precipitator, in: Proceedings of the 17th Conference of Electrostatic Precipitation, Chinese Society of Electrostatic Precipitation, 2017.
[22] Q.Z. Zheng, Application Research on Fine Particle Matters Control and Emission Characteristics of ESP and WFGD With Coal-Fired Boilers, Zhejiang University, 2017.
[23] S.L. Wang, Y. Chen, P. Han, B. Xu, Z.Z. Li, Z.W. Guo, Q.Z. Zheng, X.J. Shen, S.R. Li, K.P. Yan, PM10 and PM2.5 emission control by electrostatic precipitator (ESP) for coal-fired power plants III: applications with A 4 × 330 MW power plant, Sci. Technol. Rev. 32 (33) (2014) 39–42.
[24] B. Huang, Q. Yao, S.R. Li, Progress in technology of electrostatic enhancement for removal of PM2.5, Power System Eng. 19 (6) (2003) 44–46.
[25] B. Huang, Q. Yao, Q. Song, Z.W. Long, S.Q. Li, Effect of the electrostatics on the fibrous filter filtrating fly ash, Proc. CSEE 26 (24) (2006) 106–110.

[26] J.M. Stanley, Y. Zhuang, E.C. Michael, R.O. Michelle, Advance Hybrid Particulate Collector Final Topic Report For Phase III, U.S. Department of Energy Office of Scientific and Technical Information, 2000.
[27] K.Z. Ohtsuk, M. Shimoda, H. Asano, Mechanism of fabric filtration by electrostatic augmentation, J. Electrost. 18 (1) (1986) 93–102.
[28] C.F. Sun, W.T. Zhang, Z.H. Peng, H.H. Zheng, Z.Q. Su, Application of testing technology for fine particles of large-sale electrostatic fabric integrated precipitator, Guangdong Electric Power 28 (1) (2015) 12–15.
[29] K.X. Chen, Removal of multi-pollutants from coal-fired flue gas by electrostatic-fabric integrated precipitator based on field tests, Res. Environ. Sci. 30 (6) (2017) 937–942.
[30] Z.F. Zhao, Q. Du, G.B. Zhao, Field experimental research on PM2.5 generation and emission characteristics of pulverized coal and CFB boilers in power plants, CIESC J. 66 (3) (2015) 1163–1170.
[31] H.L. Yue, Y.T. Yuan, Z. Hong, Analysis of dust removal mechanism in sprayer of limestone/gypsum wet flue gas desulfurization system, Electric Power Environ. Protect. 22 (6) (2006) 13–15.
[32] H.G. Wei, W.P. Ye, L. Chai, Analysis of dust removal effect in wet-FGD system and its efficiency-enhancement techniques, Electric Power 48 (8) (2015) 33–36.
[33] H. Wang, Q. Song, Q. Yao, Experimental study on removal effect of wet flue gas desulfurization system on fine particles from a coal-fired power plant, Proc. CSEE 28 (5) (2008) 1–7.
[34] D.G. Wang, F.H. Zhu, Y.P. Yi, Removal efficiency of WFGD system to particles based on field tests, Admin. Technol. Environ. Monitor. 27 (5) (2015) 21–24.
[35] D.P. Pan, H. Wu, J.J. Bao, Removal effect of wet flue gas desulfurization system on fine particles and SO_3 acid mist from coal-fired power plants, Proc. CSEE 36 (16) (2016) 4356–4362.
[36] Ministry of Environmental Protection of the People's Republic of China, Practical Technical Guidelines for Pollution Prevention and Control in Coal-fired Power Plants (HJ2301-2017), 2017.
[37] X.L. Sun, F.H. Zhu, S. Wang, H. Li, Methodologies on choosing appropriate technical roadmap for ultra-low PM emission from coal-fired power plants, J. Environ. Eng. Technol. 8 (2) (2018) 129–136.
[38] F.H. Zhu, Y.Y. Xu, S. Wang, Reviewing the ultra-low emissions technology major progress of coal-fired power plants and analyzing its application effect, Environ. Prot. 44 (6) (2016) 59–63.
[39] X.L. Li, D.B. Zhou, J.X. Duan, W.J. Zhang, J.Z. Li, J.B. Yan, Characteristics of particle emission from coal-fired power plants under the ultra-low emission standard, Environ. Monitor. China 34 (3) (2018) 45–50.
[40] H.W. Chen, Z.W. Liang, B. Zhao, Experimental research on fine particles emission characteristics of a ultra-low emission coal-fired unit, Thermal Power Generation 46 (11) (2017) 108–113.
[41] R.C. Staehle, R.J. Triscori, K.S. Kumar, Wet Electrostatic Precipitators for High Efficiency Control of Fine Particulates and Sulfuric Acid Mist, ICAC Technology Forum, Institute of Clean Air Companies, Nashville, Tennessee, 2003.
[42] Z. Li, Y. Zhang, Y. Zhu, The effect of fine particulate matter on the ultra-low flue gas emission transformation of coal-fired units, Electric Power Technol. Environ. Protect. (1) (2018) 27–31.
[43] H.X. Liu, Y.P. Yao, J.G. Li, Research of PM removal mechanism and emission characteristics for WESP in coal-fired power plant, Electric Power 50 (12) (2017) 178–184.

[44] Z.D. Yang, C.H. Zheng, S.J. Liu, Y.S. Guo, C.S. Liang, Y. Wang, D.Q. Hu, X. Gao, A combined wet electrostatic precipitator for efficiently eliminating fine particle penetration, Fuel Process. Technol. 180 (2018) 122–129.
[45] D.Y.H. Pui, S.C. Chen, Z. Zuo, $PM_{2.5}$ in China: measurements, sources, visibility and health effects, and mitigation, Particuology 13 (2014) 1–26.
[46] R. Altman, G. Offen, W. Buckley, I. Ray, Wet electrostatic precipitation demonstrating precipitation demonstrating promise for fine particulate control—part II, Power Eng. 105 (2) (2001) 37–39.

CHAPTER SIX

Trace elements characteristics of ultra-low emission coal-fired power plants

Yongsheng Zhang
School of Energy, Power and Mechanical Engineering, North China Electric Power University, Beijing, China

Coal is a mixture of various substances; in addition to the main elements of C, H, O, N, S, etc., it also contains a variety of metallic and nonmetallic elements. Some of the harmful trace elements (TEs) have attracted more attention. For example, harmful metals in the NESHAP (National Emission Standards for Hazardous Air Pollutants) developed by the United States in 2015 for hazardous air pollutants in coal-fired power plants included Mercury (Hg), Antimony (Sb), Arsenic (As), Beryllium (Be), Cadmium (Cd), Chromium (Cr), Cobalt (Co), Lead (Pb), Manganese (Mn), Nickel (Ni), Selenium (Se), and so on.

Coal-fired power plants are the important sources of anthropogenic atmospheric releases of hazardous trace elements (HTE) because a large quantity of emissions can cause wide dispersion. Tian summarized and gave a multiple-year comprehensive emission inventory of the, including Hg, As, Se, Pb, Cd, Cr, Ni, and Sb for the periods 2000–2010 [1]. Because of the cobenefit removal effects of traditional PM/SO_2/NO_x control devices, emissions of these trace elements have shown a gradual decline since the peak in 2006. The total emissions in China of Hg, As, Se, Pb, Cd, Cr, Ni, and Sb are substantial and are estimated at about 118.54, 335.45, 459.4, 705.45, 13.34, 505.03, 446.42, and 82.33 tons (t), respectively, in 2010.

The global mercury assessment report of the United Nations Environment Programme (UNEP) in 2013 also shows that increasing industrialization has made Asia the largest source of mercury emissions, and that human mercury emissions in East and Southeast Asia account for about 40% of global anthropogenic mercury emissions [2]. In 2011, China issued the emission standards for air pollutants from coal-fired power plants. In this document, the emission of mercury and its compounds was included for the first time, with an emission limit of 0.03 mg/m^3. However, no emission limits were specified for other heavy metals.

Advances in Ultra-low Emission Control Technologies for Coal-Fired Power Plants
https://doi.org/10.1016/B978-0-08-102418-8.00006-1

Based on the vaporization behavior, TEs are classified into three groups: class I, class II, and class III, in which class I the trace elements are equally distributed between bottom ash and fly ash, class III the trace elements are volatized and emitted fully in the vapor phase and not enriched on the fly ash [3].

The TEs not vaporized during combustion will compose the matrix of fly ash or bottom ash. The elements partially or fully vaporized will undergo additional transformations and partitioning downstream, as the flue gas cools down in the upper furnace, which is the heat recovery section of the combustion system. The TE conversions are decided by three complex and interrelated processes: adsorption, condensation, and chemical transformation [4].

Linak and Wendt summarized TE size distribution in particles and concluded that most regulations concerned TEs, which are enriched in submicron particles [5]. Part of the vapor phase fraction is achieved through cooling of the flue gas and collection of the fine particulate. Thus, some particulate removal technologies will be beneficial for removal of these trace metals.

As for arsenic, the WFGD system is considered to have high collection efficiency for gaseous arsenic. WFGD systems had a high removal ability for arsenic with 75% to over 90% of the gaseous arsenic being absorbed by the gypsum slurry [6–9].

The selenium chemistry in the WFGD systems is highly complex and not completely understood. In the WFGD system, Se particulate dissolves in the aqueous phase of the absorbent slurry, while gaseous Se diffuses first through the gas to the aqueous phase where it dissolves and undergoes a series of chemical reactions. However, the presence and form of various components in an actual WFGD system depends on the operating conditions [10].

Based on the data before 2010, Tian summarized the cobenefit removal efficiencies from different types of flue gas control devices as shown in Table 6.1 [1]. It was found that there are high TEs removal efficiency for the traditional PM/SO_2/NO_x control devices.

The most volatile TEs (Hg, Se, As, etc.), especially mercury, are easily released into the air and are paid close attention to. In the following discussion, we will give a section, which will focus on the control of mercury by APCDs cobenefit. Several cases of mercury, arsenic and lead emissions from ultra-low emission power plants are also introduced.

Table 6.1 Release ratio and cobenefit removal efficiency of TEs from coal-fired power plants [1]

	Category	Hg	As	Se	Pb	Cd	Cr	Ni	Sb
Release ratio, R (%)[a,b]	Pulverized-coal boiler	99.42	98.46	96.22	96.25	94.94	84.50	56.53	89.40
	Fluidized-bed furnace	98.92	75.60	98.05	77.33	91.50	81.33	68.36	74.38
	Stoker-fired boiler	83.15	77.18	80.95	40.10	42.53	26.74	10.46	53.52
Coremoval efficiency, η (%)[b]	ESP	33.17	86.20	73.78	97.17	96.47	98.54	92.92	83.46
	FFs	67.92	99.00	65.00	99.00	97.63	95.13	94.83	94.25
	Cyclone	6.00	43.00	40.00	12.10	22.90	30.05	39.93	40.00
	WFGD	57.22	80.38	74.87	78.40	80.50	86.00	80.00	82.10
	SCR	11.92							
	ACI	90.00							

[a]Release ratio is comparing with the total amount of the corresponding hazardous TE in feed coal.
[b]Differences of release ratio and cobenefit removal among provinces are neglected since they are mainly technology related.

6.1 APCDs cobenefit on the mercury control

There is still much work that needs to be done in the control of mercury pollutants in China. For example, the online monitoring facilities of power plants are not complete, resulting in the lack of basic emission information, and thus the emissions are not clear. Although the emission of mercury and its compounds is limited by GB-13223-2011, the corresponding control measures and technology are still not efficient enough to capture mercury. Mercury pollution control technology is rarely used in the field. Mercury removal from coal-fired power plants has just begun. Activated carbon adsorption and mercury removal technology has been tested in Shenhua Guohua Sanhe Power Plant. The modified fly ash adsorption and mercury removal technology developed by North China Electric Power University has been applied in Shenhua Guohua Sanhe Power Plant and Shenhua Guohua Xuzhou Power Plant.

Three main forms of mercury are found in the flue gas of power plants: elemental mercury (Hg^0), oxidized mercury (Hg^{2+}), and particulate mercury (Hg^P) [11]. At present, there are three main ways to control mercury emission: coal cleaning, enhanced mercury oxidation in the furnace, and mercury removal in flue gas [12–14]. There are two main types of control technologies for mercury removal in flue gas. One is captured as a cobenefit of the existing PM, NO_x, and SO_2 controls, which mainly use existing denitrification equipment to oxidize mercury, using dust removal and desulfurization facilities to remove the mercury and its compounds. The other is by using activated carbon, fly ash, and other adsorbents to spray into the flue gas to adsorb and remove the mercury and its compounds.

The ultra-low emission technology is mainly an upgrade or to adopt a new technology to strengthen denitrification, dust removal, and desulfurization. This chapter focuses on the analysis of these technologies and the cobenefit control of their combined effects on mercury.

6.1.1 Cobenefit control of mercury in denitrification facilities

At present, the denitration technology used in power plants mainly includes SNCR and SCR. Because SCR is widely used in power plants, the cobenefit control effect of SCR on mercury is analyzed here. The main chemical reaction mechanism of SCR is that NH_3 selectively reduces NO in flue gas to N_2 under certain temperature and catalyst, and produces water at the same time. The results show that SCR has an important influence on the form and emission of mercury in flue gas, and Hg^0 can be oxidized to Hg^{2+} to facilitate its removal in the desulfurization absorber [15].

NH_3 acts as a reducing agent in the SCR and cannot oxidize Hg^0. The HCl in flue gas can oxidize Hg^0 to Hg^{2+} under the condition of catalyst [16]. The reaction equation is as follows [17].

$$Hg^0 + 2HCl + 1/2O_2 \rightarrow HgCl_2 + H_2O \tag{6.1}$$

The content of chlorine in coal is positively correlated with the content of mercury oxide in SCR, and the content of HCl in flue gas, catalyst, and quality of coal combustion all affect the ability of mercury oxide. It should be pointed out that mercury is oxidized in SCR to facilitate the removal of mercury in subsequent equipment, and there is not much effect in the direct removal of mercury. Table 6.2 shows the distribution of mercury in the flue gas before and after the SCR denitration process of four units in a power plant [18].

Table 6.2 shows that the concentration of mercury in flue gas is basically unchanged before and after denitrification, indicating that the effect of mercury reduction was not obvious, but after SCR, the concentration of mercury in different valence states had changed. It can be seen from the table that the concentration of Hg^0 decreased after SCR, while both Hg^{2+} and Hg^P increased. This indicates that SCR denitrification equipment can promote the transformation of Hg^0 to Hg^{2+}. Since Hg^{2+} is soluble in water, it will improve the removal efficiency of mercury in downstream desulfurization facilities while also having a good cobenefit control effect on mercury in flue gas. According to this mechanism, SCR cobenefit control technology for mercury can be started from the following aspects: (1) fuel optimization and adjustment, coal combustion in the furnace should produce enough halogen to oxidize mercury in flue gas; in actual running, it can be achieved by blending high and low halogen coal; (2) operation optimization and adjustment, in the SCR operation, the ammonia-nitrogen ratio needs to be strictly

Table 6.2 Distribution of mercury concentration in flue gas before and after SCR

	SCR inlet (μg/m³)				SCR outlet (μg/m³)				
Sample	**Hg^T**	**Hg^0**	**Hg^{2+}**	**Hg^P**	**Hg^T**	**Hg^0**	**Hg^{2+}**	**Hg^P**	**Mercury oxidation ratio (%)**
Unit 1	28.69	14.06	12.17	2.46	27.86	6.94	16.48	4.44	50.64
Unit 2	35.78	16.36	15.45	3.97	33.32	9.16	19.04	5.12	44.01
Unit 3	22.60	11.69	8.18	2.73	19.46	5.73	10.06	3.67	50.98
Unit 4	31.82	16.64	13.85	1.33	30.81	2.22	26.04	2.55	86.66

controlled; otherwise, ammonia may inhibit the oxidation of mercury; (3) the development of mercury nitrate cobenefit control catalyst can simultaneously remove mercury on the basis of denitrification.

6.1.2 Cobenefit control of mercury in dust removal facilities

Most coal-fired power plants in China are equipped with fabric filters or electrostatic precipitators, and the dust removal efficiency is generally above 99% [19]. The electrostatic precipitator can also remove Hg^P in particulate form in flue gas while removing dust. The fabric filter can not only remove particulate mercury, but also has the certain ability of oxidizing and adsorbing elemental mercury.

The dust in the electrostatic precipitator passes through the charged motor and enters the dust collecting area. The charged dust moves in the direction of the dust collecting plate under the action of the electric field. The moving direction of the dust is perpendicular to the direction of the flue gas flow, so that the dust particles in the process easily collide with the mercury in the flue gas, so that the mercury in the flue gas adsorbs on the surface of the particles, thereby achieving the effect of mercury removal.

In the fabric filter, when the flue gas passes through the filter cloth, the pore size of the filter cloth is small, and the dust is filtered by the filter cloth, and the fine dust is mainly collected by diffusion and sieving. The cobenefit control of mercury in the fabric filter is better, mainly because the dust layer formed on the bag has a higher ability to adsorb Hg^0 and Hg^P; on the other hand, the metal oxide contained in the dust layer converts Hg^0 in the flue gas into Hg^{2+}; the residence time of the flue gas in the fabric filter is long, which increases the contact time of the fly ash and the flue gas. Meanwhile, the probability of Hg^0 oxidation is increased.

Xu Yueyang tested the coal-fired unit that did not implement ultra-low emission. The research found that most Hg^P could be removed by ESP. Because fly ash could oxidize part of Hg^0 into Hg^{2+}, Hg^{2+} did not change significantly in ESP [20].

The process of the ESP and WESP can be divided into three phases, namely, charging, collecting, and cleaning the fly ash. Different from the ESP, the WESP collects the fly ash by using the water to rinse the collecting plate instead of rapping.

The electrode of WESP could product corona electronic, and some free radicals like O, OH, O_3 could be generated from O_2, H_2O of the flue gas as reactions (6.1)–(6.4). The higher voltage could promote the gas discharge of

O_2 and H_2O in flue gas to form more reactive free radicals. These free radicals are highly reactive toward other substances. The generation of active free radicals promoted the rapid oxidation of Hg^0 to HgO as reactions (6.5) and (6.6) or other Hg^{2+}. The Hg^{2+} in flue gas is water soluble, which could be dissolved in the water film over the dust-collecting plate of the WESP. That is, Hg^0 could be oxidized into Hg^{2+} by free radicals, and the formed Hg^{2+} could be subsequently removed by WESP. The reactions in the WESP may be as follows [21–23]:

$$e^* + O_2 \rightarrow +2{\bullet}O \tag{6.2}$$

$$O + O_2 + M \rightarrow e + {\bullet}OH + H \tag{6.3}$$

$$H_2O + e^* \rightarrow e + {\bullet}OH + H \tag{6.4}$$

$$Hg^0 + {\bullet}O \rightarrow HgO \tag{6.5}$$

$$Hg^0 + {\bullet}O_3 \rightarrow HgO + O_2 \tag{6.6}$$

6.1.3 Cobenefit control of mercury in desulfurization facilities

At present, the main desulfurization methods used in power plants are wet, dry, and semidry flue gas desulfurization technologies. In view of the wet flue gas desulfurization (WFGD) technology used in most power plants, this chapter mainly uses WFGD to illustrate the cooperative control of mercury in desulfurization facilities.

Most of the oxidized mercury compounds in the flue gas, such as $HgCl_2$, are soluble in water, and the desulfurization system captures it by dissolving the oxidized mercury in the flue gas.

Due to the high solubility of $HgCl_2$ in water, wet FGD offers the possible side benefit of removing $HgCl_2$ from the flue gas efficiently. Therefore, wet FGD systems have considerable influence on mercury removal [24]. Formation of Hg^{2+} and Cl^- complexes in FGD slurries [25]:

$$HgCl_2 + (x-2)Cl^- \leftrightarrow [HgCl_x]^{(2-x)} \quad 0 \le x \le 4 \tag{6.7}$$

When the gaseous compound of Hg^{2+} is absorbed by the liquid in the wet flue gas desulfurization system, the dissolved substance chemically reacts with the dissolved sulfide (such as hydrogen sulfide) in the flue gas to form mercury sulfide, which is stained in the liquid solvent. The mud form appears. As mentioned earlier, WFGD has the effect of removing particulate matter, so it can also remove part of the particulate mercury.

The high mercury removal efficiencies (on a coal-feed basis) in FGD units will be in the range from 50% to 75% [26, 27]. However, the proportion of mercury in flue gas from different kinds of coal combustion is different. For example, the proportion of oxidized mercury in flue gas from bituminous coal combustion is higher than that from lignite and subbituminous coal, and the mercury removal efficiency of the desulfurization device is higher than that of lignite and subbituminous coal. The mercury removal efficiency of bituminous coal flue gas is only 5%–40% if the power plant only has FGD. When the units are preceded by SCR, devices that enhance the oxidation of Hg^0 to the Hg^{2+}, the range increases to 85%–90% [26, 27]. Depending on the FGD process, a portion of this Hg may be incorporated into the FGD slurry and its solid byproducts, including synthetic gypsum.

Table 6.3 shows the mercury forms in coal-fired flue gas before and after the flue gas desulfurization equipment of five power plants. The data show that the proportion of Hg^{2+} in the flue gas of Unit 1 and 2 power plants decreased after WFGD, and the proportion of Hg^0 increased significantly. However, the proportion of Hg^{2+} and Hg^0 in the flue gas of Unit 3, 4, and Unit 5 power plants after WFGD had not changed significantly. The cobenefit removal technology of mercury in flue gas was very different, and the mercury removal efficiency was 23%–79%.

It has also been found that a small part of Hg^{2+} in WFGD was reduced to Hg^0, which is called re-emission. This may be related to hydrogen sulfate ions and iron, cobalt and manganese ions in the desulfurization slurry. Higher concentration of SO_2 or other S (IV) can promote the reduction reaction. In the absence of sufficient sulfide in the liquid solvent, the mercury sulfide reacts with the sulfite and reduces the Hg^{2+} to zero-valent mercury [30]. Operating parameters in the desulfurization system, such as pH, Cl^- content, temperature, and transition metals, all affect the reduction and re-emission of Hg^{2+}. Cheng et al. systematically studied the reaction between Hg^{2+} and S (IV) in WFGD, and obtained the following reaction mechanism [31]:

$$Hg^{2+} + SO_3^{2-} \leftrightarrow Hg(SO_3)^- \tag{6.8}$$

$$Hg(SO_3)^- + SO_3^{2-} \leftrightarrow Hg(SO_3)_2^{2-} \tag{6.9}$$

$$Hg(SO_3)_2^{2-} \rightarrow HgSO_3(s) + SO_3^{2-} \overset{H_2O}{\rightarrow} Hg^0 + HSO_4^- + H^+ + SO_3^{2-} \tag{6.10}$$

$$HSO_3^- + Hg^{2+} + H_2O(l) \leftrightarrow Hg^0 + SO_4^{2-} + 3H^+ \tag{6.11}$$

Table 6.3 Effect of WFGD on the distribution of mercury in flue gas [28, 29]

Unit	Removal efficiency (%)	Concentration before WFGD (μg/Nm^3)	Distribution before WFGD (%)			Concentration after WFGD (μg/Nm^3)	Distribution after WFGD (%)		
			Hg^0	Hg^{2+}	Hg^P		Hg^0	Hg^{2+}	Hg^P
1 [28]	40.58	18.95	68.54	30.50	0.92	11.26	71.67	27.71	0.62
2 [28]	31.74	15.03	31.62	66.04	2.36	10.26	43.76	53.69	2.50
3 [29]	43.87	8.07	50.55	49.44	0.00	4.53	81.68	18.54	0.00
4 [29]	79.22	24.35	26.45	73.51	0.00	5.06	91.10	8.89	0.00
5 [29]	22.79	2.94	85.03	14.28	0.00	2.27	93.38	6.16	0.00

Formula (6.10) is liable to occur at low sulfite concentration. In order to prevent the captured Hg^{2+} from being reduced to Hg^0, the measures to be taken are to add the trapping agent to the desulfurization slurry, so that Hg^{2+} can form stable complexes or precipitates. The commonly used inhibitors are Na_2S_4 and NaHS.

6.1.4 Cobenefit control of mercury by APCDs

In summary, denitrification, dedusting, and desulfurization equipment all play an important role in the cobenefit control of mercury. Dedusting equipment can remove Hg^P, desulfurization equipment can remove Hg^{2+}, Hg^0 is the most difficult to remove, but can be oxidized to Hg^{2+} in the SCR device, the increase of the proportion of Hg^{2+} in flue gas indirectly enhances the absorption rate of mercury by desulfurization slurry, thereby improving the whole process of mercury removal efficiency in flue gas. The absorption rate improves the overall mercury removal efficiency of mercury in the flue gas. The most important step in optimizing cobenefit mercury removal is to maximize the amount of oxidized mercury. The mercury removal rate of the wet flue gas desulfurization system is up to 90%, and the maximum oxidation rate of the selective catalytic reduction reaction can reach more than 85%, thereby improving the mercury capture rate of the wet flue gas desulfurization. Of course, re-emission should be avoided in wet flue gas desulfurization. Srivastava et al. summarized the impact of existing different pollutant control facilities on the cobenefit removal of mercury, as shown in Table 6.4.

Xu Yueyang conducted field test analysis and statistics on 20 coal-fired power plant units that did not implement ultra-low emissions [20]. The cobenefit of various pollutant control facilities on the removal of mercury from coal-fired flue gas is shown in Table 6.5. It can be found from the table that the average mercury removal efficiency of China's coal-fired units was in a large range, the lowest was only 11.4%, and the highest was 80.8%, although different APCDs combinations were used before the implementation of ultra-low emissions.

The following is a detailed discussion about the TEs through some cases in ultra-low emission power plants.

6.2 Mercury characteristics before and after ultra-low emission upgraded based WESP roadmap

6.2.1 Sampling method

The system under study is 300-MW subcritical natural circulation coal-fired unit named Unit 4 Sanhe Power Generation Co Ltd. The pollutant control

Table 6.4 Pollution controls and their connection to mercury capture [27]

System	Effect on mercury capture
CS-ESP	Good capture of particulate- or sorbent-bound mercury; better native capture for bituminous coals than low-rank coals
HS-ESP	Low native capture; probably requires specially formulated sorbents for high-temperature mercury capture
FF	Good cobenefit capture; Hg0 may be oxidized across the FF
CS-ESP + wet FGD	Good native capture for bituminous coals because of presence of soluble Hg2 + in the flue gas; relatively poor capture for low-rank coals
HS-ESP + wet FGD	Moderate native capture for bituminous coals; poor native capture for low-rank coals
FF + dry scrubber	Very high native capture expected for bituminous coals, less for low-rank coals
FF + wet FGD	Good cobenefit capture for bituminous coals; Hg0 may be oxidized across the FF and captured in the wet scrubber
SCR + CS-ESP	Good capture of particulate- or sorbent-bound mercury and better native capture for bituminous coals than low-rank coals are expected
SCR + HS-ESP	Low native capture is expected
SCR + CS-ESP + wet FGD	Good capture of particulate- or sorbent-bound mercury; better native capture for bituminous coals than low-rank coals; SC R will tend to enhance capture for bituminous coals by oxidizing Hg0 to the Hg2 + form
SCR + spray dryer + FF	Very high native capture for bituminous coals, less for low-rank coals; SC R may enhance capture by oxidizing Hg0 to Hg2 + form
SCR + HS-ESP + wet FGD	Poor capture of particulate-bound mercury in general and total mercury for low-rank coals; SC R may enhance capture for bituminous coals by oxidizing Hg0 to the Hg2 + form
SCR + FF + wet FGD	High level of mercury capture would be expected for all coals. SC R may enhance capture for bituminous coals by oxidizing Hg0 to the more soluble Hg2 + form

APC, air pollution control; *CAIR*, Clean Air Interstate Rule; *CAMR*, Clean Air Mercury Rule; CCRs, coal combustion residues; CS-ESP, cold-side electrostatic precipitator; *ESP*, electrostatic precipitator; *FF*, fabric filter; *FGD*, flue-gas desulfurization; *HS-ESP*, hot-side electrostatic precipitator; *LSFO*, limestone-forced oxidation; *MEL*, magnesium-enhanced lime; *PAC*, powdered activated carbon; *PM*, particulate matter; *SCR*, selective catalytic reduction; *SDA*, spray dryer absorber; *UBC*, unburned carbon.

unit before ultra-low emission upgrade includes low nitrogen burners, SCR denitrification reactor for NO_2 removal, electrostatic precipitator for dust removal, and wet desulfurization absorber for SO_2 removal. The low nitrogen burner reduced the fuel-type NO_x mainly by controlling the redox

Table 6.5 Mercury collaborative control of APCDs [20]

APCDs	Average mercury removal efficiency (%)	Number of sampling units
PC + ESP + WFGD	57.4 (11.4~80.8)	11
PC + FF + WFGD	67.1 (55.9~79.3)	3
PC + SCR + ESP + WFGD	70.0 (48.3~86.1)	4
CFB + Calcium injection in furnace + ESP	69.3 (62.2~76.4)	2

atmosphere in the boiler, and the electrostatic precipitator reduced the dust concentration in the flue gas by absorbing and collecting charged dust particles. The wet desulfurization used the reaction of limestone and SO_2 in the flue gas to produce gypsum. In addition, an air preheater was installed in the flue between the SCR reactor and the ESP. The system before ultra-low emission upgrading is shown in Fig. 6.1. The reformation of a low nitrogen burner implemented into the ultra-low emission upgrade included adding the exhaust air, using the bias burner 1 to further improve the oxidation and reduction conditions of the furnace. The new low-temperature economizer was mainly used to reduce flue gas temperature and improve the efficiency of the electrostatic precipitator. The improvement of a high-frequency power supply of ESP was implemented to improve the efficiency of the ESP. The desulfurization efficiency of wet desulfurization was improved, and the tubular mister was increased. A rigid plate wet ESP

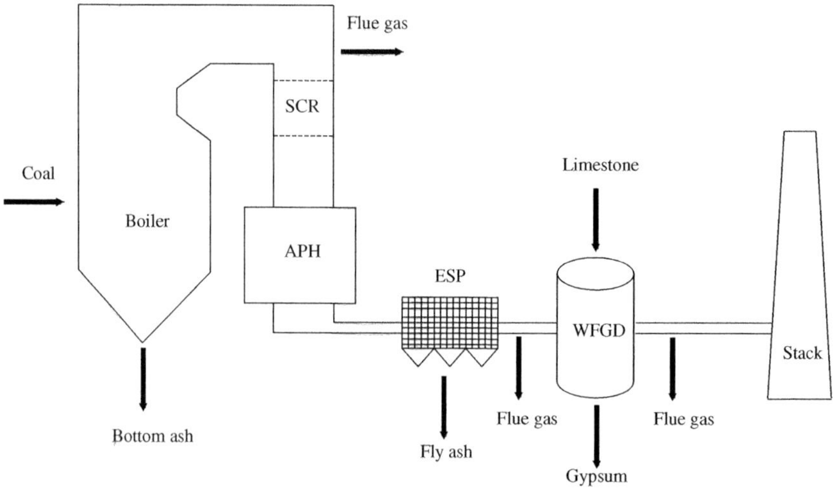

Fig. 6.1 APCDs and sampling locations before ultra-low emission upgrading.

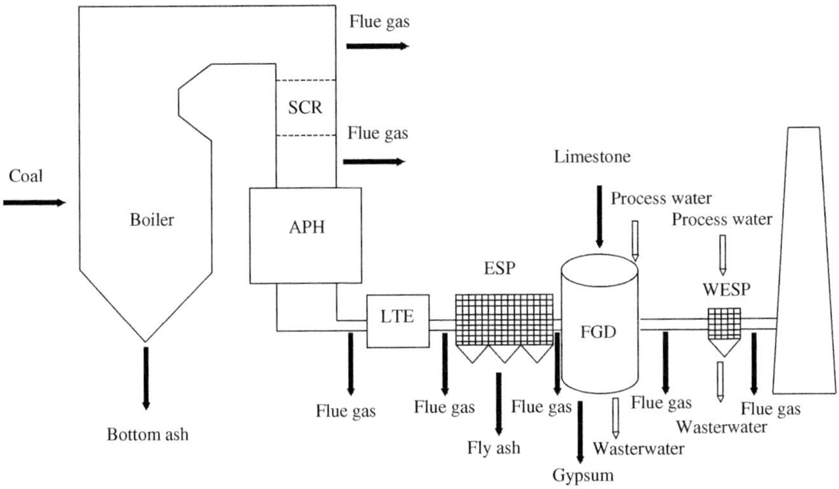

Fig. 6.2 APCDs and sampling locations after ultra-low emission upgraded.

was added to remove fine particles. The system after ultra-low emission upgrade is shown in Fig. 6.2.

Before the ultra-low emission was upgraded, flue gas samples were collected in August 2013, sampling locations included the ESP inlet, the ESP outlet, and the FGD outlet. Solid and liquid samples included coal samples from the coal feeder, dust samples from the dust collector, slag samples from the boiler slag discharging machine, gypsum samples from the desulfurization absorber, process water samples (water samples from the WFGD inlet), desulfurization wastewater samples and limestone samples, as shown in Fig. 6.1. In December 2015, after the ultra-low emission was upgraded, flue gas samples were taken again, sampling locations included the SCR inlet, the SCR outlet, the low-temperature economizer inlet, the ESP inlet, the ESP outlet, the FGD outlet, and the WESP outlet. Solid and liquid samples included coal samples from the coal feeder, dust samples from the dust collector, slag samples from the boiler slag, gypsum samples from the desulfurization absorber, process water samples (water samples from the WFGD outlet, WESP outlet), desulfurization wastewater samples, the system is shown in Fig. 6.2. In order to ensure data consistency, three sets of 30B samplers were sampled at the same time before the ultra-low emission upgrading. After the ultra-low emission upgrading, four sets of samplers were sampled at the same time before the SCR, after the SCR, before the low temperature economizer and before the ESP, and then the four sets of samplers were sampled at the same time before the ESP, after the ESP, after the FGD, and after the WESP.

6.2.2 Equilibrium calculation

As a trace element, errors in sampling and measurement analysis for mercury of the coal-fired unit are an important issue. In order to investigate the accuracy and reliability of the final data, material balance is usually used to verify the data. At the same time, the equilibrium calculation is also helpful to understand the direction and distribution of mercury in coal. It is based on the balance calculation of material input and output in the power plant system. The main formula is as follows:

$$\mathrm{RI} = \frac{M_{\mathrm{outlet}}}{M_{\mathrm{inlet}}} \tag{6.12}$$

In Eq. (6.12), the RI (Recovery Index) is the ratio of an output value to an input value of mercury in coal-fired power plants, which is called equilibrium coefficient. When mercury is in complete equilibrium in a power plant, the value of equilibrium coefficient is equal to 1 [32]. Generally speaking, field-monitoring equilibrium coefficients for heavy metals, such as mercury, are acceptable between 0.7 and 1.3 [33].

In theory, M_{inlet} is the total amount of mercury in the input material of the power plant system. Before the ultra-low emission upgrading, it includes the amount of mercury in coal, limestone, and other solids, and the amount of mercury in the process water imported from the desulfurization absorber. After the ultra-low emission was upgraded, the mercury content in the inlet process water of wet electrostatic precipitation should also be included, as shown in Formula (6.13). M_{outlet} is the total amount of mercury in the output material of the power plant system. Theoretically, before the ultra-low emission upgrading, it includes the total amount of mercury in flue gas, slag, ash, gypsum, desulfurization wastewater, and other substances. After the ultra-low emission was upgraded, it also includes the amount of mercury in the outlet process water of wet electrostatic precipitation, as shown in Eq. (6.14). In balance calculation， The M_{inlet} and M_{outlet} of each substance are the product of its mass and mercury concentration. The mercury concentration data of different substances are obtained by sampling and analysis. The quality data can be calculated, based on the actual operation running data of power plants or coal quality analysis.

$$M_{\mathrm{inlet}} = M_{\mathrm{coal}} + M_{\mathrm{limestone}} + M_{\mathrm{FGD-inlet}} + M_{\mathrm{WESP-inlet}} \tag{6.13}$$

$$M_{\mathrm{outlet}} = M_{\mathrm{flue\ gas}} + M_{\mathrm{slag}} + M_{\mathrm{fly\ ash}} + M_{\mathrm{gypsum}} + M_{\mathrm{FGD-outlet}} + M_{\mathrm{WESP-outlet}} \tag{6.14}$$

6.2.3 Mercury emissions and distribution characteristics before ultra-Low emissions upgrading

Ultimate and proximate analysis of coal before ultra-low emission upgrading are shown in Table 6.6, and mercury content in coal, ash, slag, limestone, and gypsum is shown in Table 6.7. The data show that the mercury content in coal was 49 ug/kg. Compared with the Chinese coal in literature [34], it is found to belong to low mercury content coal. After combustion, mercury was enriched in fly ash particles, and the mercury content in fly ash was about twice as much as that in coal. Because mercury in coal is released in the high-temperature combustion stage, mercury content in slag was very low for this sampling. Although the mercury content in limestone was lower than that of coal, the mercury content in the gypsum desulfurization was about 500 μg/kg due to the absorption of oxidized mercury in flue gas by the liquid phase in the wet desulfurization process.

In flue gas, the mercury concentration before the SCR was 5.02 $\mu g/m^3$, the mercury concentration after the ESP was 3.56 $\mu g/m^3$, the mercury concentration after the FGD was 1.87 $\mu g/m^3$, as shown in Fig. 6.3. From the SCR, pollutant control units such as denitrification, dust removal, and desulfurization led to a 62.7% reduction of mercury concentration. Although the 30B sampling method used in this study cannot monitor particulate mercury in flue gas, the particulate w(Hg) in flue gas is relatively small, such as particulate mercury less than 2% reported in literature [35], especially after desulfurization, most mercury in flue gas is elemental mercury. In general, the mercury emission value of the power plant was much lower than the mercury emission limit of the coal-fired power plant stipulated in GB 13223-2011 due to the low mercury content in coal resulting in low mercury concentration in flue gas.

Table 6.6 Ultimate and proximate analysis before ultra-low emission upgrading

Ultimate analysis (w/%)					Proximate analysis (w/%)			
C	H	N	S	O	Volatile matter	Ash	Fixed carbon	Moisture
64.34	4.06	0.84	0.42	14.93	27.69	15.41	50.78	6.12

Table 6.7 Mercury content of solid samples before ultra-low emission upgrading

Sample	Coal	Ash	Slag	Gypsum	Limestone
(μg/kg)	49	97	2	509	17

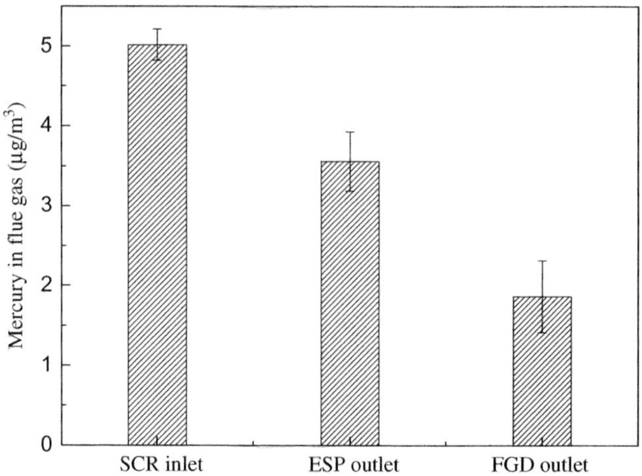

Fig. 6.3 Mercury concentration of flue gas before ultra-low emission upgrading.

Table 6.8 Mercury distribution and balance before ultra-low emission upgrading

Classification	Project	w/%
Inlet	Coal	99.6
	Limestone	0.4
Outlet	Ash	35.0
	Slag	0.1
	Gypsum	29.5
	Flue gas	35.4
Recovery index		0.73

Based on the monitoring of mercury in solid and flue gas, the mercury balance calculation and mercury distribution were carried out for the unit before the ultra-low emission upgrading. The results are shown in Table 6.8. It can be seen that mercury in the system mainly comes from coal combustion, and limestone accounts for a small part of the input. After coal burning and passing through the pollutant control units, 35.0% of the mercury existed in ash, 29.5% of the mercury existed in gypsum, and 35.4% of the mercury was discharged from the stack.

6.2.4 Mercury emission and distribution characteristics after ultra-low emission upgraded

After ultra-low emission upgraded, the ultimate and proximate analysis of coal is shown in Table 6.9. The mercury content in solid samples such as coal, lime, slag, limestone, and gypsum is shown in Table 6.10. The mercury

Table 6.9 Ultimate and proximate analysis after ultra-low emission upgraded

Ultimate analysis (w/%)					Proximate analysis (w/%)			
C	H	N	S	O	Volatile matter	Ash	Fixed carbon	Moisture
66.35	4.56	1.12	0.70	16.19	26.77	10.96	49.09	13.18

Table 6.10 Mercury content of solid samples after ultra-low emission upgraded

Sample	Coal	Ash	Slag	Gypsum	Limestone
(μg/kg)	30	147	3	601	3

Table 6.11 Mercury content of liquid samples after ultra-low emission upgraded

Liquid sample	FGD-inlet water	FGD-outlet water	WESP-inlet water	WESP-outlet water
(μg/L)	0.078	0.153	0.098	0.099

content in coal and limestone was lower in this test, 30 and 3 μg/kg, respectively, but the mercury content in fly ash was higher, about five times as much as the mercury content in coal. Mercury content in liquid samples is shown in Table 6.11; the mercury content in water increased a few times after the desulfurization absorber, the mercury in the desulfurization wastewater was 0.153 μg/L, while The mercury content in the inlet and outlet water of the wet electrostatic precipitation was basically the same, 0.098 and 0.099 μg/L, respectively.

After the ultra-low emission upgraded, the mercury concentration in the flue gas is shown in Fig. 6.4. Because of the lower content in coal, the mercury concentration in the flue gas at the SCR inlet was only 3.31 μg/L in this experiment, and the mercury concentration discharged into the atmosphere after wet electrostatic precipitation was 0.46 μg/L. Beginning from the SCR, the mercury reduced by 86.1% through the pollutant control units and the heat exchange units such as catalytic denitrification reactor, air preheater, low-temperature economizer, electrostatic precipitator, wet desulfurization absorber, wet electrostatic precipitator. The reduction of elemental mercury in the denitrification reaction absorber, the air preheater, and the low-temperature economizer was obvious, and its reduction in the electrostatic precipitator was little. After the electrostatic precipitator, mercury concentration in flue gas was 0.71 g/m^3. After the desulfurization

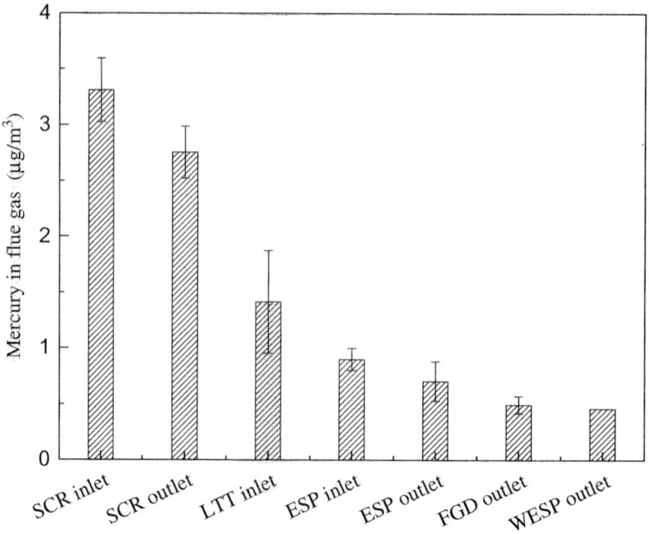

Fig. 6.4 Mercury concentration of flue gas after ultra-low emission upgraded.

absorber, the mercury concentration in flue gas was 0.50 g/m^3. The wet electrostatic precipitator had no effect on the removal of mercury in this test.

The air preheater was installed between the SCR and the low-temperature economizer. Fig. 6.5 shows that the air preheater has strong

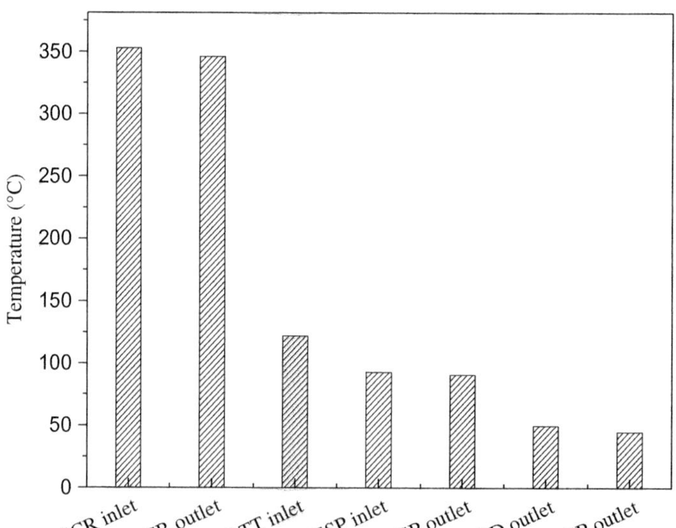

Fig. 6.5 Temperatures of different locations after ultra-low emission upgraded.

heat transfer capacity. The flue gas temperature from the outlet of the SCR to the inlet of the low temperature economizer decreased by about 220°C. At the same time, the temperature of the low-temperature economizer has also decreased to a certain extent, and the decrease of temperature is conducive to promoting the physical adsorption of mercury by fly ash in flue gas [36, 37]. After ultra-low emission upgrade, the catalytic oxidation ability of the SCR will be enhanced, which will promote the oxidation of mercury, but on the other hand, the catalyst will also promote the conversion of SO_2 to SO_3 [38]. The fly ash in the flue gas also will have the ability to adsorb SO_3, and the adsorption of SO_3 and mercury in the fly ash has a competitive relationship [39]. It is possible that the adsorption capacity of fly ash to mercury was reduced. The change of mercury concentration in flue gas is a comprehensive reflection of various factors.

Based on the monitoring data of mercury in solid, liquid, and flue gas, the mercury balance calculation and the mercury distribution analysis were carried out for the upgraded units with ultra-low emission. The results are shown in Table 6.12. Similarly, mercury in the flue gas system mainly comes from coal combustion. The influent water of desulfurization and the wet electrostatic precipitator had less effect on the mercury balance. After coal combustion and a pollutant control unit, 36.1% of the mercury existed in ash, 55.2% of the mercury was in gypsum, and 8.7% of mercury was discharged from the stack. Compared with the distribution of mercury before and after the ultra-low emission upgraded, it can be seen that the proportion of mercury in fly ash remains slightly changed, while the proportion of mercury in gypsum increased. The result of ultra-low emission upgraded enhanced the catalytic oxidation ability of the SCR to mercury and further

Table 6.12 Mercury distribution and balance after ultra-low emission upgraded

Classification	Item	w/%
Inlet	Coal	98.2
	Limestone	1.8
	FGD-inlet water	<0.01
	WESP-inlet water	<0.01
Outlet	Ash	36.1
	Slag	0.1
	Gypsum	55.2
	FGD-outlet water	<0.01
	WESP-outlet water	<0.01
	Flue gas	8.7
Recovery index		1.14

absorbed mercury in the desulfurization absorber. The reason for the less effect on the Hg adsorption in the ESP system may be due to the increase of SO_3 in the flue gas to inhibit the physical adsorption of oxidized mercury in fly ash.

6.2.5 Mercury emissions from wastewater of FGD and WESP

The mercury content in wastewater from ultra-low emission power plants is also a concerning issue for academia and industry. To better understand the mercury transformation in the water system from ultra-low emission, within this section the characteristics of mercury in wastewater from the 300-MW coal-fired unit of Sanhe Power Plant was studied.

(1) Flue gas desulfurization system

The flue gas desulfurization system includes absorber, spray layer, tube bundle-type mist eliminator, slurry circulation pump (As shown in Fig. 6.6). The flue gas enters from the bottom of the spray layer of the absorber and was in countercurrent contact with the circulating slurry. The SO_2 in the flue gas reacted with the $CaCO_3$ in the limestone slurry (As shown in Table 6.13) at the upper absorption zone of the absorber to form calcium sulfite. The calcium sulfite was oxidized to calcium sulfate by the air, which blows in by the oxidation fan at the lower pool of the absorber. Finally, the calcium sulfate was sent to the gypsum dehydration system through the gypsum discharge pump. The molar ratio of calcium to sulfur in the desulfurization system was less than 1.03. The limestone

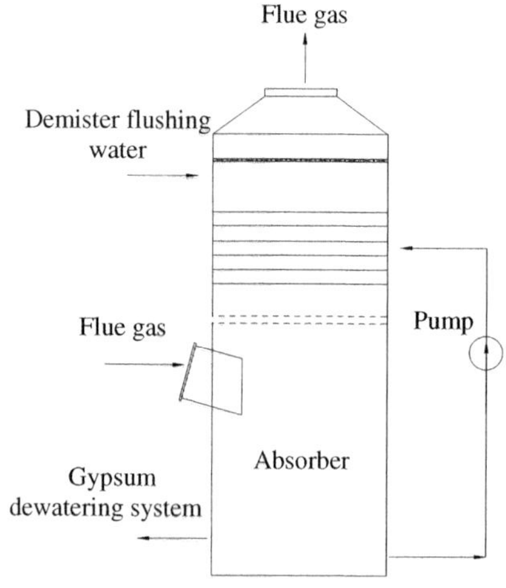

Fig. 6.6 Sketch map of desulfurization system.

Table 6.13 Limestone contents (w/%)

Name	Content	Name	Content
$CaCO_3$	93.84	Fe_2O_3	0.074
Na_2O	0.04	K_2O	0.036
SiO_2	0.33	MnO	<0.005
Al_2O_3	0.22	H_2O	0.02

dosage was 2.9 t/h when the boiler was running at full load and the gypsum output of desulfurization by-product was 5.06 t/h.

According to the process requirements, the desulfurization system needs to continuously discharge a certain amount of wastewater to maintain the concentration of Cl^- in the absorber. The total amount of evaporated flue gas and wastewater discharged from the desulfurization system as rated under load was about 52 m^3/h. In order to maintain the water balance of the desulfurization system, it is necessary to replenish the desulfurization system with process water. The water source was from the drainage of the circulating cooling water system of the unit, which was from the urban reclaimed water. The process water of the power plant was replenished to the desulfurization absorber water cycle by the demister flushing water.

(2) Wet electrostatic precipitator system

The wet electrostatic precipitator has been added into the ultra-low emission power plant to increase the PM captured efficiency. The WESP was a rigid plate spray-type electrostatic precipitator. The inlet flue gas temperature was designed to be 58°C. The dust removal efficiency (including gypsum) was greater than 86%. The $PM_{2.5}$ removal efficiency was greater than 86%. The wet electrostatic precipitator system is illustrated in Fig. 6.7. 12 sets of spray

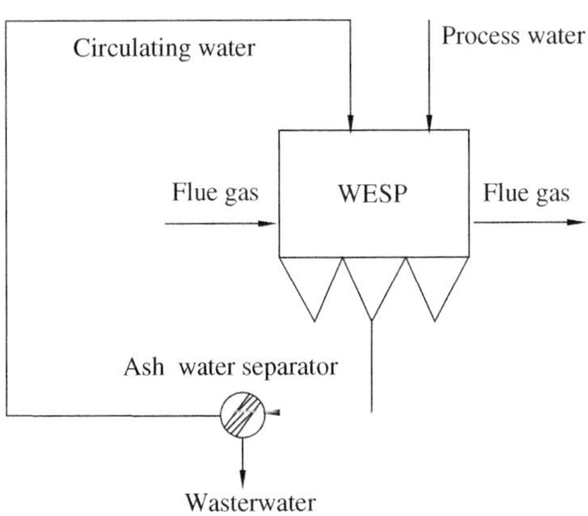

Fig. 6.7 Sketch map of wet electrostatic precipitator system.

pipes were arranged along the air flow direction. 23 nozzles were arranged on each of the tubes in the vertical air flow direction. Cathode and anode plates for continuous discharge can be flushed continuously and stably. The liquid, collected by the wet electrostatic precipitator, entered the circulating water tank. The pH of the water in the water tank was maintained in the range between 7 and 10. The pH value was adjusted by replenishing the alkali liquid. Qualified circulating water was used for hydration of the sprinkler system, while wastewater with a large concentration of suspended solids was discharged. The supplementary water source was the same as the desulfurization system and it was also the drainage of the unit circulating cooling water system. The designed wastewater discharge and process water replenishment was 3.2 t/h.

During the field test, samples involved supplement water samples from the WESP, the desulfurization wastewater samples, sewage and limestone samples from the outlet of the WESP. The experiment time was 9 days, and the unit load was about 80% of the full load during the experiment.

The mercury concentration in the desulfurization replenishing water and desulfurization wastewater is shown in Fig. 6.8. Except the mercury content in the desulfurization replenishing water sample on the first day was lower

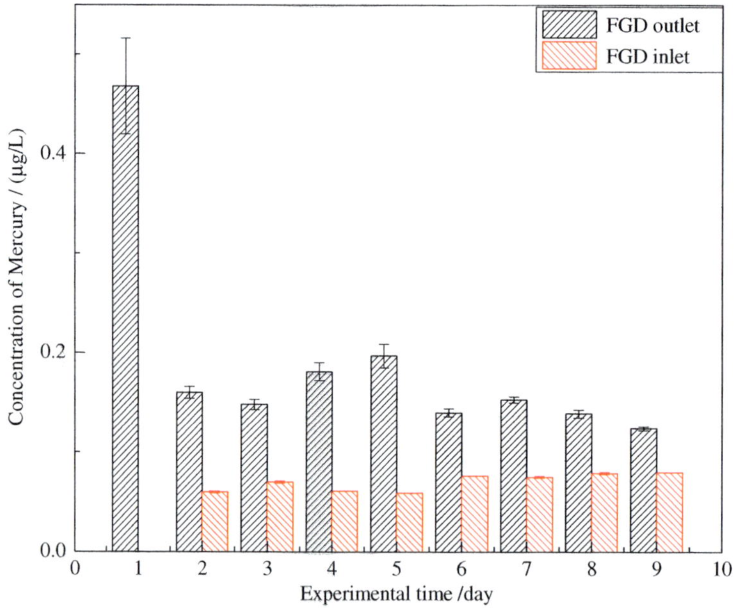

Fig. 6.8 Mercury concentration of WFGD replenishing water and wastewater.

than the instrument detection limit of 1 ng/L, the mass concentration of mercury was between 0.060 and 0.080 μg/L on other days of the samples. The mercury concentration in the first-day wastewater sample was higher than others, which reached 0.468 μg/L. In the next few days, the mercury concentration was between 0.140 and 0.197 μg/L, which is much lower than the limits on mercury emissions of the electric power unit standard in China (DL/T997-2006 "Water quality control index of limestone-gypsum wet desulfurization wastewater in thermal power plant"). According to the mercury concentration in the desulfurization replenishing water and desulfurization wastewater, the mass of mercury absorbed by waste water discharged from the desulfurization system in an hour can be calculated (As shown in Fig. 6.9, the calculation method is: 52 m^3/h × (load during experiment)/(rated load) × (mercury concentration in desulfurization wastewater − mercury concentration in replenishing water)). It can be seen from Fig. 6.9 that the increase of mercury in the desulfurization drainage on the first day was 18.6 mg/h and the increase of mercury in the desulfurization drainage in the other sampling dates was 1.75~5.50 mg/h.

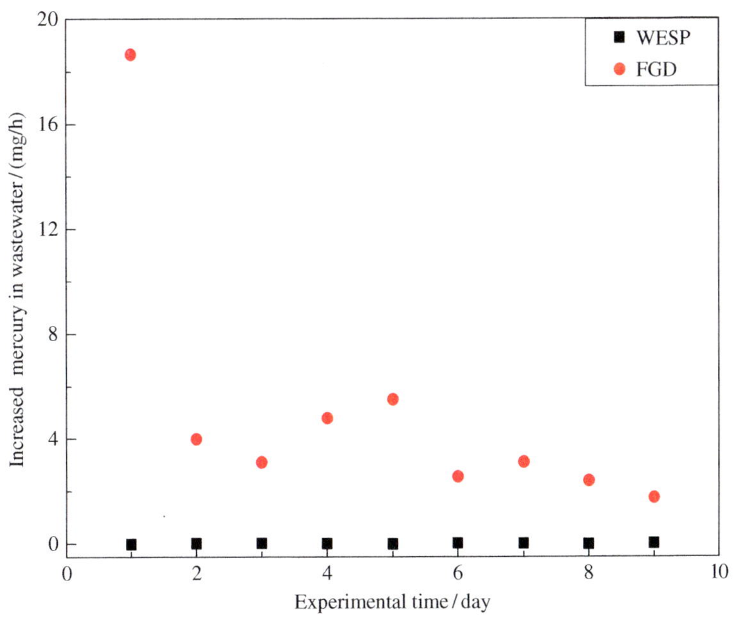

Fig. 6.9 Increased mercury in WFGD or WESP wastewater per hour.

The mass concentration of mercury in the replenishing water and sewage of the wet electrostatic precipitator was 0.096~0.111 μg/L and 0.094~0.102 μg/L, respectively, during the experiment. The mercury concentration in the liquid did not change significantly after the wet electrostatic precipitator.

The literature [40] reported a wet electrostatic precipitator experiment at the Bruce Mansfield Plant in the United States. The test bench was made of 316 stainless steel. The total mercury concentration of the inlet flue gas was 4.88 μg/m^3. The total mercury concentration of the outlet was 2.85 μg/m^3. The mercury removal efficiency of the WESP flue gas was calculated to be 41% (the particle mercury removal rate was 76%, the mercury oxide removal rate was 86%, and the elemental mercury removal rate was 18%.). In this test, the mercury concentration in the flue gas was lower than that. The mercury concentration of the WESP inlet was 0.50 μg/m^3 while the reading was 0.46 μg/m^3 for the WESP outlet. The mercury removal efficiency for the WESP was 8%.

The mercury removal effect of the WESP in this power plant was not higher than that of the other studies. The reason may be attributed to the low concentration of particulate matter after the desulfurization system under an ultra-low emission upgrade [41]; thus, the particulate mercury concentration is low. At the same time, the desulfurization system has a strong ability to absorb oxidized mercury after the ultra-low emission upgraded. Most of the mercury in the flue gas after the desulfurization is elemental mercury, and the WESP has a weaker ability to remove elemental mercury.

6.3 Mercury characteristics of ultra-low emission upgraded based on LTT-ESP roadmap

6.3.1 Sampling method

The sampling locations of solid, gas, and liquid are indicated in Fig. 6.10 for the Unit 3 set in Dagang Power Plant.

In the study, mercury samplings of flue gas were conducted before and after each flue gas pollution control unit. Due to the limitation of the number of sampling instruments, the mercury pollution field-monitoring experiments were divided into two days. Samples were taken at different monitoring positions through three sets of 30B mercury sampling devices, the first day was named Day 1, and gas samples were taken at the SCR inlet, the SCR outlet, and the ESP inlet. The next day was named Day 2, and gas

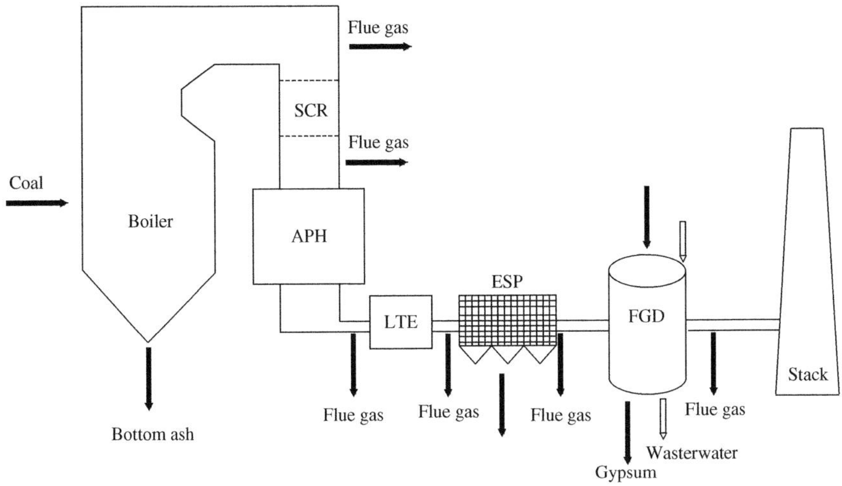

Fig. 6.10 Ultra-low Emission units and locations of Hg sampling.

samples were taken at the ESP inlet, the FGD inlet, and the FGD outlet. Among them, repeated experiments were carried out at the sampling points of the ESP inlet to investigate the stability of sampling.

The load between two days fluctuated at 280–300 MW with little change. It can be considered that the combustion conditions and the pollutant discharge of the boiler between two days were basically the same. And the results obtained at the ESP inlet were corrected to ensure the consistency of the experimental results.

During the experiment, solid and liquid samples were taken simultaneously, including coal samples from the coal feeder, ash sample from the dust collector, slag sample from the boiler slag extractor, gypsum sample from the desulfurization absorber, process water sample (water sample from the FGD inlet), desulfurization wastewater samples, and limestone samples.

6.3.2 Mercury emission and distribution characteristic

As shown in Table 6.14, the ultimate analysis and proximate analysis of coal burning in the field experiment during mercury monitoring showed that the coal quality was relatively stable. If the load fluctuated from 280 to 300 MW and the change was small during the field test, the working conditions between two days can be considered the same and the results can be compared.

The mercury content in coal, ash, slag, limestone, and gypsum during the experiment is shown in Table 6.15. The data show that the mercury content

Table 6.14 Ultimate and proximate analysis of coal

	C	H	N	S	O	Volatile matter	Ash	Fixed carbon	Moisture
Day 1	57.34	1.76	0.25	0.68	23.07	27.43	16.9	50.89	4.77
Day 2	57.82	1.89	0.21	0.75	26.42	28.98	12.91	52.87	5.25
Average	57.58	1.82	0.23	0.71	24.75	28.21	14.91	51.88	5.01

Table 6.15 Mercury content of solid sample (μg/kg)

	Coal	Ash	Slag	Limestone	Gypsum
Day 1	58	112	2.4	8.6	243
Day 2	65	150	2.7	7.2	249
Average	61.5	131	2.5	7.9	246

of coal was 61.5 μg/kg, which belongs to the coal with lower mercury content. After the pulverized coal burns in the furnace, mercury was released completely, and most of the mercury was present in the flue gas. When the mercury in flue gas flows through the subsequent flue, fly ash will adsorb it; thus, mercury accumulation occurred to a certain extent. The data show that the mercury content in fly ash was about 131 μg/kg, which was about twice as high as that in coal. The mercury content in slag generated by combustion was extremely low, mainly because the furnace temperature is high and mercury cannot exist stably in slag. The mercury concentration in limestone was low, only 7.9 μg/kg. But the mercury content in the desulfurized gypsum was about 250 μg/kg due to the absorption of oxidized mercury in flue gas by the liquid phase in the wet desulfurization process.

During flue gas monitoring, the measured concentration of mercury was converted to a concentration of 6% of the baseline oxygen content. The mercury concentration in the flue gas at different points is shown in Fig. 6.11. It can be seen that the mercury concentration in the flue gas before denitrification was 5.50 $\mu g/m^3$, and after denitrification was 5.23 $\mu g/m^3$. The denitrification system of the power plant cannot achieve the removal of mercury in the flue gas, but the denitrification system can catalyze the conversion of Hg^0 to Hg^{2+}.

The mercury concentration in flue gas from the SCR to the ESP decreased by 68%, and the mercury concentration in flue gas at the ESP inlet was 1.67 $\mu g/m^3$. The air preheater and the low-temperature economizer can effectively reduce the temperature of flue gas at the inlet of the dust collector, an air preheater, and a low-temperature economizer, which had been

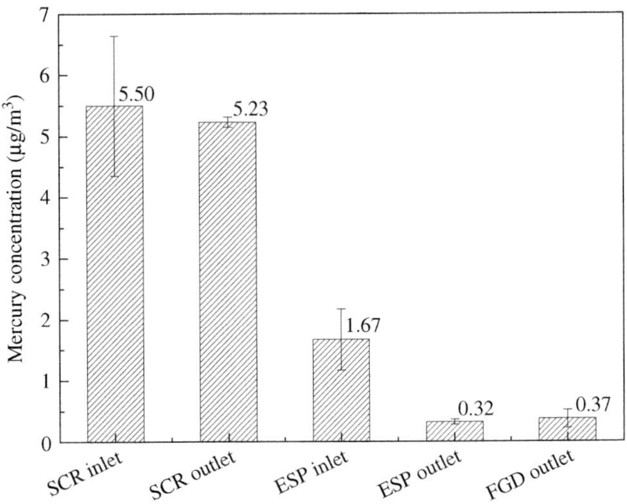

Fig. 6.11 Hg concentration in flue gas of different locations.

installed between the SCR and the ESP. The sampling temperature at each measuring point is shown in Fig. 6.12. It can be seen that the flue gas temperature from the outlet of the SCR to the inlet of the ESP was reduced by about 200°C. The temperature reduction enhanced the physical adsorption of mercury in the flue gas by fly ash and the formation of particulate mercury; thus, the mercury concentration in flue gas was reduced effectively.

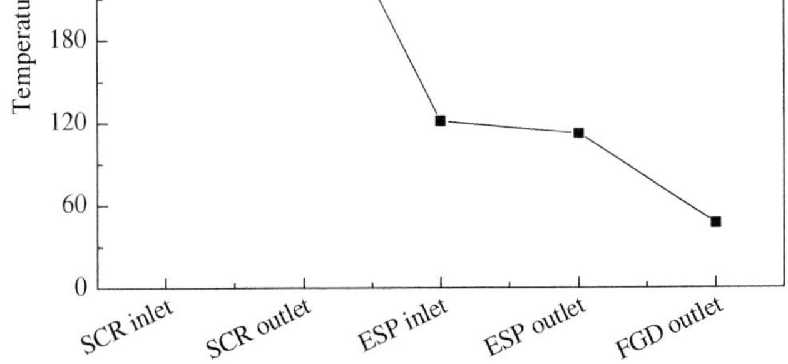

Fig. 6.12 Gas temperature of different locations.

The mercury concentration in flue gas of the ESP outlet was 0.32 μg/m^3, and the mercury concentration in flue gas of the FGD outlet was 0.37 μg/m^3. These test results show that the effect on mercury emission reduction in flue gas by the FGD is not obvious, and there is still a little re-emission. From the SCR to the FGD, the mercury concentration decreased by 93.3%, showing a higher efficiency of cobenefit mercury removal. The mercury concentration in flue gas was low due to the low mercury content in the coal, and the mercury emission of this power plant was far less than the mercury emission limit of the coal-fired power plant specified in the Emission Standard of Air Pollutants for Thermal Power Plants (GB13223-2011).

The mercury content in liquid samples such as limestone slurry, gypsum slurry, and process water is shown in Table 6.16. It can be seen that the mercury content in liquid samples is extremely low.

Based on the monitoring of mercury in solid and flue gas, the distribution of mercury in the unit after ultra-low emission upgrade is shown in Table 6.17. It can be seen that the mercury in this system mainly came from coal combustion, and the mercury in limestone accounted for a small part of the input. After the combustion of coal, the flue gas passed through the APCDs. 57.5% of mercury existed in ash, 34.1% of mercury existed in gypsum, and 8.3% of mercury was discharged from the stack. Compared with the literature [42], in which mercury in solid after combustion accounted for 61.6% of total mercury of the input, the mercury was more easily transferred to solid after the ultra-low emission ungraded.

Table 6.16 Hg concentration of liquid samples (μg/L)

	Limestone slurry	Gypsum slurry	Process water
Day1	0.30	0.40	0.17
Day2	0.32	0.35	0.12
average	0.31	0.37	0.15

Table 6.17 Hg distribution

Input (%)	Coal	99.7
	Limestone	0.3
Output (%)	Ash	57.5
	Slag	0.1
	Gypsum	34.1
	Flue gas	8.3

6.4 Arsenic emissions of different power plant units

The study compares the characteristics of arsenic emissions from eight different power plants in China and the United States. One of the power plants carried out arsenic monitoring before and after the ultra-low emission upgraded. In order to conveniently read this study, it was named Unit 2 (before the ultra-low emission upgrading) and Unit 9 (after the ultra-low emission upgraded) in this book.

6.4.1 Experimental procedure

The sampling program for the American power plant was carried out during 2003 and 2004, while the sampling time for the Chinese power plant was during 2012 and 2016. Coal, flue gas, fly ash, slag, limestone, and gypsum samples were collected from nine power plants. The APCDs for each power plant are shown in Table 6.18. Units 2 and 9 are the same unit located at the same power plant. Unit 9 upgraded with a low-temperature economizer to lower the flue gas temperature down to 90°C before entering the ESP, and a WESP was installed to enhance the capture efficiency of particulate matter to meet the ultra-low emission standards. None of the US coal-fired power plants (Units 5–8) in this study was equipped with a low-temperature economizer or the WESP system. There is only one plant with a hot ESP included in this power plant study, Unit 6.

During the arsenic sampling periods, the boiler flue gas samples were extracted from under the following operating conditions: (1) consistent load at more than 85%; (2) consistent coal supply; and (3) bringing down the coal storage level to a minimum in the coal hoppers before changing to a new coal.

Pulverized coal was collected from the pipes leading from the pulverizer to the boiler. The fly ash systems consisted of four banks ESPs in China and two banks ESPs in the US. In all cases, the first row of hoppers was sampled in China and in the US. Grab samples of slag, limestone, and gypsum were synchronized with hopper fly ash sampling. The flue gas streams were sampled at the inlet and outlet of the APCDs. The arsenic in the flue gas was sampled in accordance with EPA Method 29 "Determination of Metals Emissions from Stationary Sources." The sampling equipment consists of the Ontario isokinetic sampling instrument from American APEX Instruments. This system was composed of a sampling probe, filter, filter tank, absorption bottle package (a total of four groups, placed in an ice water bath),

Table 6.18 Electric power plant data

Unit ID	Unit 1	Unit 2	Unit 3	Unit 4	Unit 5	Unit 6	Unit 7	Unit 8	Unit 9
Capacity	200 MW	300 MW	200 MW	300 MW	300 MW	500 MW	508 MW	125 MW	300 MW
Country	CHN	CHN	CHN	CHN	US	US	US	US	CHN
Boiler	PC	PC	PC	CFB	PC	PC	PC	PC	PC
WESP	NO	NO	NO	NO	NO	NO	NO	NO	YES
SO_2 control	FGD	FGD	FGD	Injection limestone into furnace	NO	NO	FGD	NO	FGD
Particulate control	Cold ESP	Cold ESP	Cold ESP	Cold ESP	Cold ESP	Hot ESP	Cold ESP	Cold ESP	Cold ESP
Low temperature economizer	NO	NO	NO	NO	NO	NO	NO	NO	YES
NO_x control	SCR	SCR	SCR	NO	NO	NO	SCR	NO	SCR
Fuel type	Bituminous coal								

Note: PC represents pulverized coal furnace; CFB represents circulating fluidized-bed furnace; FGD represents flue gas desulfurization equipment; ESP represents electrostatic precipitator equipment; SCR represents selective catalytic reduction equipment.

and the master control box (including dry gas flow meter, vacuum gauge, pump, and ancillary equipment). In the process of sampling, the flue gas was sampled by the probe from a flue probe, where the flue gas passes through a heated quartz fiber membrane, the absorption bottle package, the control box, and a dry gas flow meter was metering the dry flue gas volume. Arsenic in the flue gas is collected by acidification of H_2O_2.Flue gas samples were collected at Units 1, 2, 5, 6, and 9.

All analyses of liquid and solid samples were performed at the North China Electric Power University and the Institute for Combustion Science and Environmental Technology, Western Kentucky University. The analyses of arsenic in the coal, fly ash, slag, limestone, and gypsum in China were performed following Chinese Method GB/T 3058-2008, using automatic hydride generation atomic fluorescence spectrometry (PSA10.055 Millennium Excalibur, PSA Company). The analyses of arsenic in the coal, fly ash, and slag in the US were performed following ASTM method D6357-OOa, using LECO renaissance inductively coupled plasma - time of flight mass spectrometry (ICP-TOFMS). Flue gas samples were analyzed using automatic hydride generation atomic fluorescence spectrometry (PSA10.055 Millennium Excalibur, PSA Company), in China and in the US, according to the sample pretreatment and analytical guidelines of EPA's Method 29. The linearity of the calibration curve for these analyses was 0.999, and the limit of detection was 0.35 μg/kg. Coal analyses were performed following standard ASTM guidelines (moisture, ash, and volatile matter—ASTM D-5142-02a; sulfur—ASTM D-4239-02a; chlorine—ASTM D-4208-02). Major oxides and minor elements were analyzed by ICP-AES (Leeman LabsProdigy) in China and by X-ray fluorescence on a Rigaku RIX 3001 XRF in the US. In order to ensure the accuracy of the arsenic test results, each group of samples was tested three times. The data were deemed valid if the relative deviation values were within ±10%. Additionally, for every five samples to be tested, a known concentration of standard liquid was used to validate instrument operational error, if the standard liquid test results were within ±10%, the test data were considered with good enough accuracy to be used in the study.

6.4.2 Emission and balance of arsenic

The proximate and ultimate data analysis of fly ash composition and arsenic concentration is shown in Table 6.19. All the US coals have high volatility (32%–37%), moderate ash content (9%–12%), and wide range of sulfur

Table 6.19 Ultimate/proximate analysis of coal samples (dry basis)

	Proximate analysis (%)			Ultimate analysis (%)				
Unit	V	A	FC	C	H	O	N	S
1#	30.74	15.67	53.58	64.91	4.36	13.77	0.86	0.43
2#	28.28	16.41	55.31	64.34	4.06	13.93	0.84	0.42
3#	25.89	28.43	45.68	60.07	4.37	6.01	0.63	0.49
4#	23.71	38.62	37.67	48.39	3.30	7.11	0.89	1.69
5#	32.12	12.59	55.30	71.49	4.94	8.41	1.32	1.25
6#	33.54	11.27	55.19	73.02	4.96	8.77	1.36	0.62
7#	37.88	10.57	51.56	65.92	5.53	13.69	1.39	2.90
8#	33.78	9.86	56.36	75.79	5.00	6.28	1.77	1.30
9#	29.31	18.48	52.21	63.43	4.21	12.24	0.89	0.75

Note: V represents volatile, A represents ash, FC represents fixed carbon, C represents carbon, H represents hydrogen, O represents oxygen, N represents Nitrogen, S represents sulfur.

contents (0.6%–2.9%). Coal of Unit 7 has a relatively high sulfur concentration (2.9%). The Chinese coals have a wide ash content range (15%–38%) with relatively low sulfur content, except the coal of Unit 4 (1.7%). As shown in Table 6.20, the major differences in the fly ash composition are among the Al, Ca, and Fe concentrations. As shown in Table 6.21, the arsenic concentration in the Chinese coals was between 0.85 and 3.34 μg/g and the arsenic concentration in the US coal was between 2.5 and 22.3 μg/g. The Arsenic concentrations in flue gas were very low at both Chinese and US plants. Only a small amount of arsenic was emitted from the stack (Unit 1: 0.21 $\mu g/m^3$, Unit 2: 0.52 $\mu g/m^3$ (flue gas after the ESP), Unit 5: 0.57 $\mu g/m^3$, Unit 6: 0.71 $\mu g/m^3$ and Unit 9: 0.10 $\mu g/m^3$). It was found that

Table 6.20 Fly ash major and minor composition analysis (%)

	Major and minor analysis											
Unit	Al_2O_3	CaO	Fe_2O_3	K_2O	Na_2O	MgO	P_2O_5	SiO_2	TiO_2	BaO	MnO	SrO
1#	31.03	2.76	2.62	2.73	0.51	0.52	0.71	50.10	1.31	0.08	0.02	0.27
2#	19.57	11.74	7.25	2.62	2.27	0.84	0.53	46.60	0.82	0.08	0.16	0.72
3#	21.09	6.24	7.92	3.75	1.55	0.11	0.87	49.87	0.79	0.13	0.11	0.38
4#	29.45	1.40	5.20	1.54	0.26	0.36	0.69	54.64	1.10	0.03	0.12	0.12
5#	19.56	2.13	7.40	2.29	0.01	0.70	0.17	51.79	1.57	0.04	0.02	0.08
6#	20.03	1.69	5.33	2.70	0.11	0.79	0.10	54.97	1.68	0.11	0.03	0.09
7#	15.56	1.12	23.17	2.16	0.21	0.75	0.07	39.66	1.00	0.03	0.02	0.03
8#	25.11	1.45	12.38	1.86	0.47	1.00	0.24	50.59	1.43	0.20	0.02	0.11
9#	18.87	12.52	7.10	2.34	2.18	0.76	0.62	48.53	0.81	0.07	0.13	0.65

Table 6.21 Arsenic concentration of all samples

Unit	Coal (μg/g)	Fly ash (μg/g)	Slag (μg/g)	Economizer ash (μg/g)	Limestone (μg/g)	Gypsum (μg/g)	Flue gas (emissions) (μg/m^3)	Flue gas (out of ESP) (μg/m^3)	Flue gas (out of FGD) (μg/m^3)
1#	1.03 ± 0.06	5.05 ± 0.03	3.11 ± 0.16	ND	1.48 ± 0.04	2.08 ± 0.07	0.21 ± 0.02	ND	0.21 ± 0.02
2#	0.85 ± 0.08	4.84 ± 0.09	1.60 ± 0.12	ND	1.12 ± 0.06	1.68 ± 0.01	ND	0.52 ± 0.04	ND
5#	11.53 ± 0.03	91.51 ± 0.25	7.50 ± 0.08	ND	ND	ND	0.57 ± 0.02	0.57 ± 0.02	ND
6#	2.53 ± 0.08	23.96 ± 0.45	5.06 ± 0.04	16.24 ± 0.12	ND	ND	0.71 ± 0.01	0.71 ± 0.01	ND
9#	1.26 ± 0.05	5.57 ± 0.24	0.83 ± 0.03	ND	1.68 ± 0.02	1.46 ± 0.06	0.10 ± 0.01	0.15 ± 0.01	0.13 ± 0.01
3#	3.05 ± 0.01	10.53 ± 0.16	2.67 ± 0.07	ND	ND	ND	ND	ND	ND
4#	3.34 ± 0.06	16.16 ± 0.34	21.27 ± 0.03	ND	ND	ND	ND	ND	ND
7#	22.25 ± 0.04	160.20 ± 0.52	12.16 ± 0.06	110.31 ± 0.32	ND	ND	ND	ND	ND
8#	4.00 ± 0.05	42.23 ± 0.21	2.86 ± 0.03	ND	ND	ND	ND	ND	ND

Note: ND represents that there is no such solid or gas samples collected.

most of the arsenic in the coal was retained in the fly ash except at the CFB in Unit 4. Generally, for a PC boiler, the arsenic content in the fly ash will be 5~10 times the arsenic content in the coal, while the arsenic content in the slag will be at the same level of that in coal. For CFB boilers, the arsenic content in the slag was higher than in fly ash, with both levels about 5 times the equivalent concentration in coal. It is known that arsenic is readily volatilized during the combustion process, after combustion in the boiler it will tend to enrich the fly and bottom ash. The fireside temperature of a PC boiler is significantly higher than that of a CFB boiler; thus, arsenic would be more condensed after combustion, and the arsenic content in the slag of a CFB would be expected to be higher. It should be noticed that for these sampled power plants, the arsenic content in limestone and gypsum was low, with concentrations between 1 and 2 mg/kg.

6.4.3 Balance of arsenic

During combustion, arsenic is released into the flue gas and gets partitioned into vapor and particulate phases. Arsenic partitioning between vapor and the particulate phase is due to the chemical interactions of arsenic with fly ash. Fig. 6.13 shows Mass balances 1 (RI 1) for arsenic in Units 2, 5, 6, and 9 have been calculated, assuming that the input was the arsenic content from the combusted coal. The particulate arsenic concentration in the filter ash collected during the flue gas sampling at the ESP inlet was used as the arsenic concentration in the fly ash. The recovery of arsenic (RI 1) was 81.92%, 88.20%, 95.20%, and 71.63% for Units 2, 5, 6, and 9, respectively.

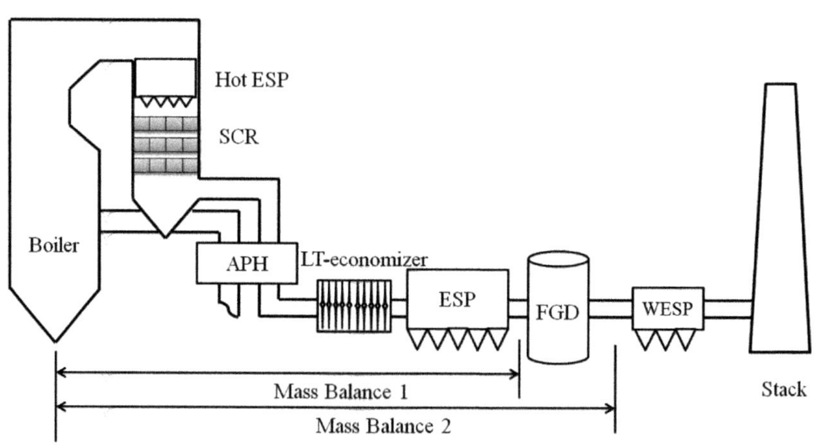

Fig. 6.13 Air pollution control devices and mass balance range.

Flue gas was only taken after the FGD in Units 1 and 9, thus, mass balance 2 (RI 2) was adopted. RI 2 for arsenic in Units 1 and 9 were calculated assuming that the input is the arsenic content from the combusted coal and limestone. Arsenic will be associated with the fly ash, slag, gypsum, and in-stack arsenic in the form of particulate and gas phases, respectively. Based on RI 2, the recovery of arsenic was 72.24% and 73.11% for Units 1 and 9, respectively. The overall mass balance for arsenic in this study was around 70%–85%. This relatively low mass balance may be due to the effect of the APCDs system as will be discussed in the following sections.

6.4.4 Effect of APCDs on arsenic removal

6.4.4.1 Effect of the SCR on arsenic removal

The recovery of arsenic is low for Units 1, 2, and 9, which are equipped with SCR systems, see Table 6.22. The arsenic recovery was high for Units 5 (without an SCR system) and 6 (with a hot ESP). The recovery index for Units 5 and 6 was 88.20% and 95.20%, respectively. Unit 9 has three catalyst layers, and Unit 2 only has two layers. The RI 1values for Units 9 and 2 were 71.63% and 81.92%, respectively. The flue gas passes through the SCR before it reached the ESP inlet. It can be thought that partial gas phase As_2O_3 solidified on both active and nonactive sites in the vanadium-based catalysts that were used in the SCR system, which reduced the activity of the catalyst. When compared with the other units, about 15% of the arsenic may be retained in the SCR system. Unit 6 has the highest arsenic recovery rate. This may be due to the location of the hot ESP, which was located before the air preheater, operating at around 350°C. Most of the arsenic compounds could be captured in the hot ESP before the flue gas enters any of the other APCD system. This explains the excellent arsenic mass balance given the layout of the APCD system. However, few power plants have a hot ESP to minimize arsenic poisoning on the catalyst in the SCR system. Most power plants select low arsenic coal to limit catalyst poisoning. It may be difficult to obtain a high recovery rate for arsenic in the SCR system.

Table 6.22 Recovery values for arsenic at different units

	Unit #1	Unit #2	Unit #5	Unit #6	Unit #9
RI 1 (%)	–	81.92	88.20	95.20	71.63
RI 2 (%)	72.24	–	–	–	73.11

6.4.4.2 Effect of the ESP on arsenic removal

In order to investigate the arsenic capturing efficiency for fly ash, the ESP fly ash samples from nine units were collected and analyzed for arsenic. The ESP arsenic capturing efficiencies were calculated. Results from Units 5 and 6 indicated that the hot ESP has a significant, positive effect on arsenic removal. This means that using hot ESPs can reduce SCR catalyst poisoning and improve the service life of the catalyst.

In this study, it was found that the particulate precipitators collect about 90% of the arsenic present. The particulate removal efficiency for the effect of the ESP units was determined by the particulate matter measurements carried out. Based on the particulate removal efficiency, the average arsenic removal efficiency of the ESP unit should be about 95%.

Units 2 and 9 are the same unit. Unit 9 upgraded with ultra-low emissions technology-based Unit 2. After the ultra-low emissions upgraded, the arsenic content in coal has increased by 50%, but the arsenic concentration decreased from 0.52 to 0.15 $\mu g/m^3$ after the ESP, decreased by 71%, showing that the ESP reduces arsenic emissions. The arsenic removal efficiency of the ESP for Units 2 and 9 was 95.87% and 98.22%, respectively. Enhancing the particle matter capture efficiency increased the arsenic removal efficiency by 2.35%. The low-temperature economizer reduced flue gas temperature at the ESP entrance to approximately 90°C. Thus, the resistivity of both the fly ash and the flow of flue gas was reduced. The ESP particulate matter removal efficiency was improved. It has been suggested that SO_x, HCl, and H_2O in the flue gas would condense and absorb on the surface of the fly ash particles and a liquid membrane would be formed on the surface of the fly ash. The conductivity and viscosity of the surface of the fly ash would also be increased. Thus, the opportunity for the arsenic compounds to be captured by the fly ash would increase. Agglomeration of smaller particles to form larger particles will occur and improve fly ash capture efficiency.

Fly ash particle size is another factor that has been measured in the literature that affects arsenic capturing efficiency. The ESP ash particle size will decrease when the flue gas flows from Row 1 to Row 4 in the ESP system. In this study, the arsenic concentrations in ashes from the first two rows of hoppers were compared. Three sets of ash samples collected from Rows 1 and 2 show a consistently larger arsenic concentration in Row 2. This trend is shown in Fig. 6.14. In general, the fly ash particle size was different in different electric fields, and the fly ash–specific surface area (SSA) was also different, and some scholars have demonstrated that fly ash SSA increases as the

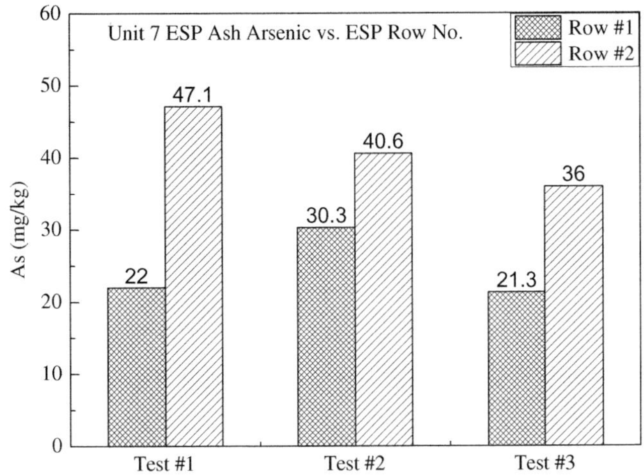

Fig. 6.14 Arsenic concentration in the ESP fly ash at the hoppers for different rows.

particle size decreases [7, 43]. The arsenic absorption capacity increased with increasing fly ash SSA. Hence, the arsenic partitioning in the ESP fly ash would be in accordance with its size; the smaller-sized particles had increased arsenic content. Although the flue gas at the ESP outlet contains only very small amounts of fine fly ash, the enrichment of arsenic in these particulates would increase the risk of contamination after the release of the ash into the environment.

6.4.4.3 Effect of FGD on arsenic removal

Limestone and gypsum samples from Units 1 and 9 were collected and analyzed for arsenic to investigate the arsenic capturing efficiency in the FGD system.

A small amount of arsenic (less than 3.49%) was collected by the FGD at Units 1 and 9. There was less arsenic than expected in the ESP fly ash, which suggests that arsenic in the particulate phase was enriched on the fine particles in the flue gas, which could escape from the ESP unit and be removed by the wet scrubber. The particulate arsenic was insoluble during the FGD process and easily incorporated into a solid portion of the slurry in the scrubber tanks.

6.4.4.4 Effect of the WESP on arsenic removal

The WESPs are used to control fine particulate as well as sulfuric acid aerosols and trace metals. It was found in this study that arsenic was more easily enriched in fine particles. In theory, fine particle removal would greatly impact arsenic removal. Arsenic concentrations in the flue gas were 0.130 and 0.098 $\mu g/m^3$at the Unit 9 WESP inlet and outlet, respectively. The arsenic removal efficiency at the WESP was 24.6%. Capturing arsenic is a cobenefit of the WESP system, but the amount of arsenic was very small in the total mass balance.

Unit 9 was upgraded with a low-temperature economizer, an ESP high-frequency power supply along with a WESP. The APCDs in Unit 9 were found to be able to remove 99.93% of the arsenic from the flue gas stream. This indicates that ultra-low emissions upgrading increased arsenic capture ability in a coal-fired power plant.

6.5 Lead emission characteristics

6.5.1 350 MW ULE unit with ESP+FF roadmap upgraded

Shilin Zhao et al. sampled a 350-MW coal-fired power plant equipped with SCR, ESP+FF, and WFGD under three loads [44]. HTEs in the flue gas at the inlet and outlet of each APCDs were sampled, based on US EPA Method 29. The content of Pb in coal was in the range of 2.94 to 8.56 mg/kg. The emission concentration was in the range of 0.09 to 0.13 mg/m^3. Most of Pb was in the particulate state, and easy to be removed by the ESP and FF, the total removal rate of lead across APCDs was over 99%. There was no obvious change for the gas phase Pb across the APCDs. More detailed information can be known from Table 6.23.

6.5.2 660-MW ULE unit with WESP roadmap

Shilin Zhao et al. sampled a 660-MW unit of ULE coal-fired power plant as shown in Table 6.24, which was equipped with SCR, ESP, WFGD, and

Table 6.23 Pb in coal and flue gas [44]

Sampling load	Coal (mg/kg)		Flue gas (mg/m^3) SCR inlet	SCR outlet	ESP+FF outlet	WFGD outlet
100%	8.56±0.27	Pbg	0.11	0.12	0.05	0.09
		Pbp	1.44×10^3	1.11×10^3	0.50	0.09
83%	4.45±0.12	Pbg	0.11	0.43	0.09	0.08
		Pbp	7.11×10^2	6.98×10^2	0.55	0.12
71%	2.94±0.16	Pbg	0.09	0.07	0.13	0.13
		Pbp	5.15×10^2	5.16×10^2	0.50	0.13

Table 6.24 Pb in flue gas (mg/m^3) [45]

	SCR inlet	SCR outlet	ESP outlet	WFGD outlet	WESP outlet
Pbg	0.98	0.59	0.04	0.47	0.16
Pbp	3.02×10^3	2.37×10^3	1.19	0.28	–

WESP [45]. Flue gas sampling of Pb was conducted by using the EPA Method 29. The content of Pb in coal was 23.4 mg/kg. The Pb emission concentration was 0.16 mg/m^3, and the Pb in flue gas after WESP was in the gaseous form. The Pb removal rate was higher than 99.90% across the APCDs, which demonstrates this ULE coal-fired power plant has excellent capture effects on TEs. In the flue gas before the ESP, particulate Pb was the main form. Thus, high dust efficient removal performance of the ESP resulted in greater Pb removal effects. The Pb removal rate across the WFGD was about 40%. More than 60% of Pb was removed by the WESP, which indicates some gaseous Pb was further removed under the conditions of more water vapor and lower temperature.

6.6 Summary

From the results of the field sampling and testing in this chapter, the implementation of the ultra-low emission upgrade also effectively reduces the emission of trace heavy metal pollutants, such as volatile mercury, arsenic, and lead.

References

[1] H.Z. Tian, K.Y. Liu, J.R. Zhou, L. Lu, J.M. Hao, P.P. Qiu, J.J. Gao, C.Y. Zhu, K. Wang, S.B. Hua, Atmospheric emission inventory of hazardous trace elements from china's coal-fired power plants temporal trends and spatial variation characteristics, Environ. Sci. Technol. 48 (2014) 3575–3582.

[2] UNEP, Global Mercury Assessment 2013: Sources, Emissions, Releases and Environmental Transport, UNEP Chemicals Branch, Geneva, Switzerland, 2013.

[3] J.A. Ratafia-Brown, Overview of trace elements partitioning in flames and furnaces of utility coal-fired boilers, Fuel Process. Technol. 39 (2) (1994) 139–157.

[4] M.H. Xu, R. Yan, C.G. Zheng, Y. Qiao, J. Han, C.D. Sheng, Status of trace element emission in a coal combustion process: a review, Fuel Process. Technol. 85 (2003) 215–237.

[5] W.P. Linak, J.O.L. Wendt, Trace metal transformation mechanisms during coal combustion, Fuel Process. Technol. 39 (2) (1994) 173–198.

[6] C.B. Wang, H.M. Liu, Y. Zhang, C. Zou, E.J. Anthony, Review of arsenic behavior during coal combustion: volatilization, transformation, emission and removal technologies, Prog. Energy Combust. Sci. 68 (2018) 1–28.

[7] R. Meij, Trace element behavior in coal-fired power plants, Fuel Process. Technol. 39 (1–2) (1994) 199–217.
[8] P. Córdoba, R. Ochoa-Gonzalez, O. Font, M. Izquierdo, X. Querol, C. Leiva, M.A. López-Antón, M. Díaz-Somoano, M.R. Martinez-Tarazona, C. Fernandez, A. Tomás, Partitioning of trace inorganic elements in a coal-fired power plant equipped with a wet flue gas desulphurisation system, Fuel 92 (1) (2012) 145–157.
[9] E. Álvarez-Ayuso, X. Querol, A. Tomás, Environmental impact of a coal combustion-desulphurisation plant: abatement capacity of desulphurisation process and environmental characterisation of combustion by-products, Chemosphere 65 (11) (2006) 2009–2017.
[10] P. Córdoba, Partitioning and speciation of selenium in wet limestone flue gas desulphurisation systems: a review, Fuel 202 (2017) 184–195.
[11] H.C. Hsi, C.Y. Tsai, T.H. Kuo, Development of low-concentration mercury adsorbents from biohydrogen-generation agricultural residues using sulfur impregnation, Bioresour. Technol. 102 (16) (2011) 7470–7477.
[12] H.P. John, A.S. Everett, D.M. Michael, Status review of mercury control options for coal-fired power plants, Fuel Process. Technol. 82 (2003) 89–165.
[13] N. Stephen, V.N. Chitralkumar, S.B. Mark, M. Larry, Interpreting enhanced Hg oxidation with Br addition at Plant Miller, Fuel Process. Technol. 90 (11) (2009) 1372–1377.
[14] Y.J. Zheng, A.D. Jensen, C. Windelin, Review of technologies for mercury removal from flue gas from cement production processes, Prog. Energy Combust. Sci. 38 (2012) 599–629.
[15] J. Xiang, J.H. Qiu, Y.H. Xiong, Experimental study on NOx emission characteristics of boilers, Proc. CSEE 20 (9) (2000) 80–83.
[16] L. Xu, M. Mei, C.T. Li, A. Shazadan, et al., Research progress on mercury pollution control of coal-fired flue gas using existing equipment and technology, Environ. Eng. 28 (3) (2010) 77–80.
[17] H. Kamata, S.I. Ueno, T. Naito, et al., Mercury oxidation by hydrochloric acid over a VOx/TiO_2, catalyst, Catal. Commun. 9 (14) (2008) 2441–2444.
[18] F.S. Liu, Y.J. Xia, R. Xu, et al., Mercury removal effect and mercury emission characteristics test of pollution control equipment in coal-fired power plants, China Electric Power 50 (4) (2017) 162–166.
[19] P. Lian, L. Zhu, A.M. Fang, et al., Study on the application and selection of dust collectors in coal-fired power plants, Power Sci. Technol. Environ. Protect. 27 (2) (2011) 18–21.
[20] Y.Y. Xu, J.M. Xue, H.L. Wang, et al., Study on synergistic control of mercury in conventional pollutant purification facilities for coal-fired flue gas, Proc. CSEE 34 (23) (2014) 3924–3931.
[21] Y. Zhang, J.P. Yang, X.H. Yu, P. Sun, Y.C. Zhao, J.Y. Zhang, G. Chen, H. Yao, C.G. Zheng, Migration and emission characteristics of Hg in coal-fired power plant of China with ultra low emission air pollution control devices, Fuel Process. Technol. 158 (2017) 272–280.
[22] J.G. Calvert, S.E. Lindberg, Mechanism of mercury removal by O_3 and OH in the atmosphere, Atmos. Environ. 39 (2005) 3355–3367.
[23] Y. Huang, Z.C. Yin, Y. Chen, X. Guo, Experimental study on gaseous elemental mercury removal by wet electrostatic precipitators, Fuel 234 (2018) 1337–1345.
[24] I. Shigeo, Y. Takahisa, A. Kazuo, Emissions of mercury and other trace elements from coal-fired power plants in Japan, Sci. Total Environ. 368 (1) (2006) 397–402.
[25] H. Barna, H. Melanie, S. Günter, Impact of additives for enhanced sulfur dioxide removal on re-emissions of mercury in wet flue gas desulfurization, Appl. Energy 114 (2014) 485–491.

[26] C.L. Kairies, K.T. Schroeder, C.R. Cardone, Mercury in gypsum produced from flue gas desulfurization, Fuel 85 (17-18) (2006) 2530–2536.
[27] R.K. Srivastava, N. Hutson, B. Martin, F. Princiotta, J. Staudt, Control of mercury emissions from coal- fired electric utility boilers: an overview of the status of mercury control technologies, Environ. Sci. Technol. 40 (5) (2006) 1385–1393.
[28] Chinese Research Academy of Environmental Sciences, Training Course on Mercury Pollution Control Technology in Coal-Fired Power Plants, Chinese Research Academy of Environmental Sciences, Beijing, 2011.
[29] S.X. Wang, L. Zhang, G.H. Li, Y. Wu, J.M. Hao, N. Pirrone, F. Sprovieri, M.P. Ancora, Mercury emission and speciation of coal-fired power plants in China, Atmos. Chem. Phys. 10 (2010) 1183–1192.
[30] L. Chen, Y. Duan, Y. Zhou, Mercury transformation across particulate control devices in six power plants of chlorine and ash composition, Fuel 86 (4) (2007) 603–610.
[31] C.M. Cheng, Y. Cao, Z. Kai, W.P. Pan, Co-effects of sulfur dioxide load and oxidation air on mercury re-emission in forced-oxidation limestone flue gas desulfurization wet scrubber, Fuel 106 (2013) 505–511.
[32] C.M. Cheng, P. Hack, P. Chu, et al., Partitioning of mercury, arsenic, selenium, boron, and chloride in a full-scale coal combustion process equipped with selective catalytic reduction, electrostatic precipitation, and flue gas desulfurization systems, Energy Fuel 23 (10) (2009) 4805–4816.
[33] T. Yokoyama, K. Asakura, H. Matsuda, et al., Mercury emissions from a coal-fired power plant in Japan, Sci. Total Environ. 259 (1–3) (2000) 97.
[34] C.G. Zheng, Emission and Control of Mercury from Coal Combustion, Science Press, 2010.
[35] S. Tang, L. Wang, X. Feng, et al., Actual mercury speciation and mercury discharges from coal-fired power plants in Inner Mongolia, Northern China, Fuel 180 (2016) 194–204.
[36] J.C. Hower, C.L. Senior, E.M. Suuberg, et al., Mercury capture by native fly ash carbons in coal-fired power plants, Prog. Energy Combust. Sci. 36 (4) (2010) 510–529.
[37] Y.Z. Chen, F.H. Chai, Z.G. Xue, et al., Study on mercury emission factors for coal-fired power plants, Res. Environ. Sci. 19 (2) (2006) 49–52.
[38] J.A. Halstead, R. Armstrong, B. Pohlman, et al., Nonaqueous heterogeneous oxidation of sulfur dioxide, J. Phys. Chem. 94 (8) (1990) 3261–3265.
[39] Z. Ye, C. Martin, J. Pavlish, et al., Cobenefit of SO reduction on mercury capture with activated carbon in coal flue gas, Fuel 90 (10) (2011) 2998–3006.
[40] J. Reynolds, Mercury removal via wet ESP, Power 148 (8) (2004).
[41] S.M. Wang, Y. Zhang, J.Z. Liu, Integrated application of fine particle control technology in coal-fired power plants and "near zero emission" characteristics, Environ. Sci. Res. 29 (9) (2016) 1256–1263.
[42] M.L. Hui, L. Zhang, Z.G. Wang, et al., The mercury mass flow and emission of coal-fired power plants in China, China Environ. Sci. 35 (8) (2015) 2241–2250.
[43] X. Guo, C.G. Zheng, M.H. Xu, Characterization of arsenic emissions from a coal-fired power plant, Energy Fuel 18 (2004) 1822–1826.
[44] S.L. Zhao, Y.F. Duan, L. Chen, Y.N. Li, T. Yao, S. Liu, M. Liu, J.H. Lv, Study on emission of hazardous trace elements in a 350MWcoal-fired power plant. Part 2. arsenic, chromium, barium, manganese, lead, Environ. Pollut. 226 (2017) 404–411.
[45] S.L. Zhao, Y.F. Duan, H.Z. Tan, M. Liu, X.B. Wang, L.T. Wu, C.P. Wang, J.H. Lv, T. Yao, M. She, H.J. Tang, Migration and emission characteristics of trace elements in a 660 MW coal-fired power plant of China, Energy Fuel 30 (2016) 5937–5944.

CHAPTER SEVEN

Cost effectiveness of ultra-low emission for coal-fired power plants

Yongsheng Zhang
School of Energy, Power and Mechanical Engineering, North China Electric Power University, Beijing, China

Cost effectiveness of ultra-low emission for coal-fired power plants is one of the key issues drawing public concern. Zhao et al. adopted a cost-effective evaluation method [1], took into account factors such as coal quality and technology, and based their analysis upon the presumption that the unit emission reduction costs for WFGD, SCR, and ESP were 154, 69, 15 CNY/(kW a), respectively. Also, with the electricity cost being used as an indicating index, a study baseline was obtained. Through this, the electrical cost of the ultra-low emission upgrade was estimated at about $2.94{\sim}3.59$ (10^{-2} CNY/kW h). Moreover, they believed the ultra-low emission upgrade cost would decrease with the increase in the scale of equipment configuration, improvement of technical level, and purification of coal quality. It was estimated that the cost of ultra-low emission upgrade in China accounted for about 0.24% of its 2014 domestic GDP.

The mentioned study was based on macroanalysis. In actual ultra-low emission upgrade practice, the actual situation faced by various coal-fired power units differ, and the upgrade plans and the project quantity were all different. Thus, the cost difference would also be quite significant. This chapter will discuss the issues through the case study involving specific power plants.

YinBiao Chen et al. analyzed the cost effectiveness of two 300-MW and two 350-MW coal-fired power units having underwent ultra-low emission upgrade [2, 3], as shown in Table 7.1, with fixed investment of $50{\sim}70$ million CNY.

Xu Hao made an estimation about the extra added investment cost concerning the ultra-low system upgrade for 300-, 660-, and 1000-MW coal-fired power units (see Table 7.2 for parameters before and after the ultra-low emission upgrade) [4]. They were about 254.4, 189.4, and 163 CNY/kW

Advances in Ultra-low Emission Control Technologies for Coal-Fired Power Plants
https://doi.org/10.1016/B978-0-08-102418-8.00007-3

Table 7.1 Investment of four upgraded ultra-low emission units in the same coal-fired power plant

	Capacity (MW)	Ultra-low emission upgraded roadmap	Investments on fixed assets (million CNY)
Unit 1	350	LNB + SCR + LTE + ESP (High-frequency power supply) + WFGD + WESP (flexible collectors)	58.56
Unit 2	350	LNB + SCR + LTE + ESP (High-frequency power supply) + WFGD + WESP(rigid collectors)	61.98
Unit 3	300	LNB + SCR + LTE + ESP (High-frequency power supply) + WFGD	49.55
Unit 4	300	LNB + SCR + LTE + ESP (High-frequency power supply) + WFGD (rigid collectors)	70.37

Table 7.2 Basic conditions for technical and economic analysis [4]

Emission parameters		330 MW	660 MW	1000 MW	APCDs
Before ULE upgrading	Flue gas flow (m^3/h)	1,150,000	2,070,000	3,210,000	
	PM (mg/Nm^3)	30	30	30	ESP
	NO_X (mg/Nm^3)	100	100	100	Low nitrogen burners + SCR
	SO_2 (mg/Nm^3)	200	200	200	FGD
After ULE upgrading	PM (mg/Nm^3)	5	5	5	LTE ESP, high-frequency power supply, WESP
	NO_X (mg/Nm^3)	50	50	50	Optimized ammonia injection, increased catalyst
	SO_2 (mg/Nm^3)	35	35	35	Added spray layer, pulp pools outside the absorber

respectively. As to the coal-fired power units undergoing ultra-low emission upgrade, it was found that the gross investment increased along with the increase of unit capacity, while the investment per unit decreased along with the increase of unit capacity. For these units, the annual operating costs newly added were approximately 17 million CNY, 27 million CNY, and 38.9 million, respectively. The operating costs newly added per unit were approximately 1.4 (10^{-2} CNY/kW h), 1.2 (10^{-2} CNY/kW h), and 1.0 (10^{-2} CNY/kW h), as shown in Table 7.3 details.

Wang Shumin et al. analyzed the ultra-low emission upgrade costs involving 300-, 600-, 800-, and 1000-MW coal-fired power units. As a consequence, they were of the opinion that the investment operation cost for ultra-low emission power plants would increase by $0.54 \sim 1.13 \times 10^{-2}$ - CNY/kW h [5]; more detail as Table 7.4 shows.

Xiong Yuehui's study shows that electricity generation cost per unit for 300-MW and 600-MW coal-fired power plants would increase by about 2×10^{-2} and 1.5×10^{-2} CNY/kW h, respectively, after the ultra-low emission upgrade [6]. Zhang Donghui et al. calculations show that if the ultra-low emission upgrade for newly built 600-MW coal-fired power units were carried out, the investment needed would be between 345 and 439 CNY/kW. For the newly constructed 300-, 600-, 1000-MW coal-fired power units

Table 7.3 Estimated investment of ultra-low emission upgrade for coal-fired power units (Unit: million CNY) [4]

Cost	330 MW	660 MW	1000 MW
Engineering construction	17.2	21.6	25.4
Equipment purchase	36.7	62.7	81.7
Project installation	21.5	30.7	38.5
Others	8.6	10.0	17.4
Total	84.0	125.0	163.0
Unit kW investment CNY/kW	254.5	189.4	163
Increased annual operating	17.0	27.0	38.9
Increased operating (10^{-2} CNY/kW h)	1.41	1.20	1.01

Table 7.4 Cost statistics of the near-zero emissions power plants of Shenhua Group [5]

Coal-fired power unit	Capacity (MW)	Increased operating electrical prices (10^{-2} CNY/kW h)
Sanhe Power Plant Unit 4	300	1.125
Dingzhou Power Plant Unit 1	600	1.120
Cangdong Power Plant Unit 3	660	1.060
Suizhong Power Plant Unit 1	800	1.068
Suizhong Power Plant Unit 3	1000	0.542

having achieved the ultra-low emission, electricity generation cost per unit would increase by 1.87×10^{-2}, 1.43×10^{-2}, and 0.926×10^{-2} CNY/kW h, respectively [7].

Cheng et al. investigated the data of 129 ultra-low emission power units using different technology roadmaps of LTT-ESP, WESP, and ESP-FF [8]. The technology roadmap details as adopted by these units were shown in Fig. 7.1. It was obvious that LTT-ESP and WESP PM removal equipment were in greater application.

Figs. 7.2 and 7.3 show the statistical data describing both the investment and operating cost per unit of those power units under statistics. Investment cost per unit was between 150∼270 CNY/kW, and operating cost per unit was between 20∼45 CNY/kW. It was also found that the larger the capacity of the equipment unit was, the lower the investment per unit would be. Different technical roadmaps showed an irregular trend in correspondence to the power units' different capacity. And the reason might be related to specific coal types or on-site construction conditions.

Due to the differences in parameters of coal-fired power units, as well as due to the options of various technical solutions for ultra-low emissions, there would be variations regarding equipment investment and construction accordingly. This in turn would result in differences of extra operational electricity costs, caused by ultra-low emission implementation. However, it could be found that extra operating costs of ultra-low emission coal-fired power units were generally between 1×10^{-2} and 2×10^{-2} CNY/kW·h.

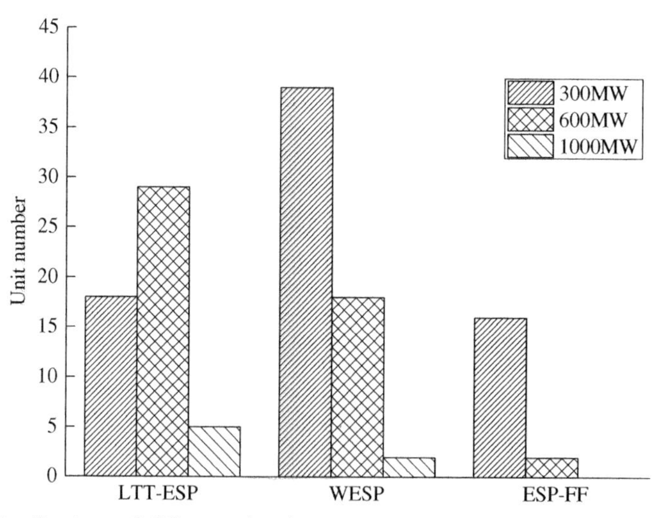

Fig. 7.1 Applications of different ultra-low emission upgrade roadmaps.

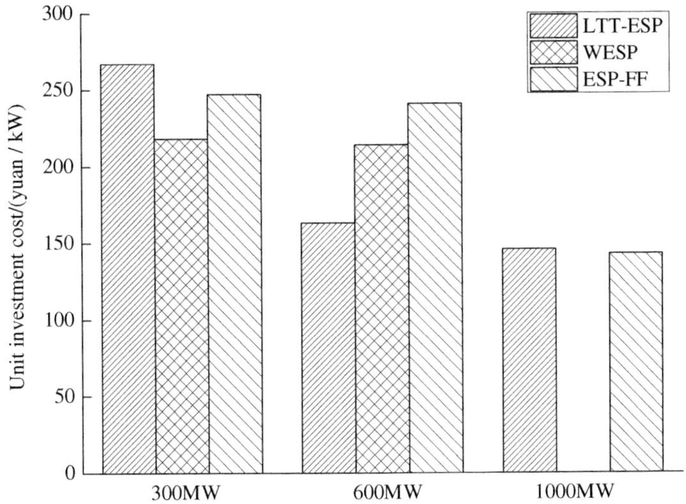

Fig. 7.2 Investment cost of different ultra-low emission upgrade roadmaps.

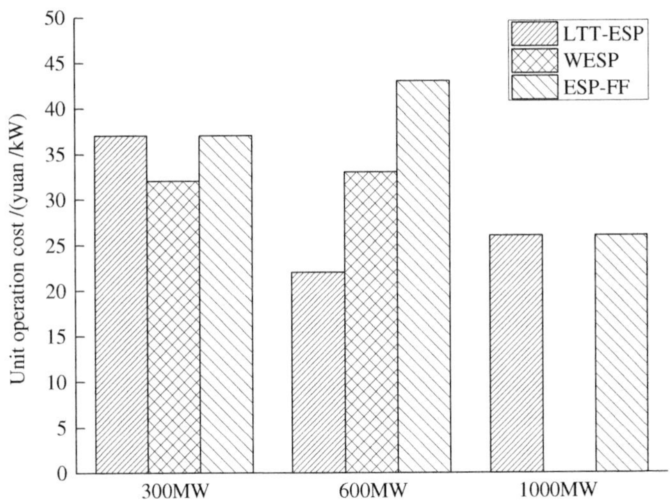

Fig. 7.3 Operating cost of different ultra-low emission upgrade roadmaps.

In the meantime, it was found that the larger the unit capacity, the lower the extra operating electricity.

Cheng compared the investment costs corresponding to specific pollutant control measures under different technology roadmaps, as adopted in the ultra-low emission upgrade for three 300-MW coal-fired power units, using LTT-ESP, WESP, and ESP-FF technical roadmaps. Among the costs, the

Table 7.5 Investment cost comparison between various technology roadmaps (Unit: million CNY)

Description	LTT-ESP	WESP	ESP-FF
SCR upgrading	13.80	13.80	13.80
LTE ESP	14.20	–	–
ESP upgrading	–	2.40	–
ESP + FF	–	–	12.12
Fan upgrading	12.50	12.50	12.50
FGD upgrading	29.18	27.65	29.18
WESP added	–	19.00	–
Total	69.68	75.35	67.60

cost of desulfurization upgrade was found to be higher than the rest. The details are shown in Table 7.5 [9]. Meanwhile, it was also found that the estimation of investment costs made by Cheng was in conformity with the actual cost of the ultra-low emission upgrade for the 300-MW coal-fired power units as shown in Table 7.1.

The following is a case study about the upgrade cost in three of the power plants as mentioned in Chapter 4.

7.1 Dagang Power Plant

In the ultra-low emission technology roadmaps deployed by the 330-MW units, Dagang Power Plant added new low-temperature economizers and installed high-efficiency demisters. It also upgraded three-phase power of ESP supply. Specific costs of this construction project are shown in Table 7.6. The newly added cost was between 64 and 67 CNY/kW, and the cost of the ultra-low emission upgrade in this power plant was relatively low.

Following the completion of the upgrade, in order to take the full advantage of the three-phase high-efficiency power supply, the power parameters were all set to full-load operational condition. Accordingly, the ESP

Table 7.6 Dagang Power Plant costs of ultra-low emission construction projects

Units	ESP upgrade (million CNY)	Newly added LTE (million CNY)	Newly added high-efficiency demisters (million CNY)	Increased cost (CNY/kW)
1, 2	19.98	18.88	3.60	64.6
3, 4	22.46	18.90	3.00	67.5

Table 7.7 Electricity consumption of ESP in Dagang Power Plant

Units	Electricity consumption (MW h/day)	Conversion to full load auxiliary power ratio (%)
1	35	0.44
2	35	0.44
3	36	0.46
4	20	0.25

efficiency was greatly improved. Among the units under operation, the electrostatic precipitator Nos. 1, 2, and 3, running under high parameters, saw an increase of power consumption by about 10,000 degrees/day, compared with use of the prior single-phase power supply. ESP power consumption of each unit after the upgrade, when converted to the plant overall power consumption rate, was between 0.25% and 0.46%. The statistical summary of the power consumption of the four ESP under operation, after upgrades, is shown in Table 7.7.

7.2 Hequ Power Plant

The ultra-low emission technology roadmaps for Units 1 and 2 of Shanxi Hequ 600-MW subcritical boiler was SCR upgrade + LTE + PM removal upgrade + desulfurization upgrade + stack anticorrosion.

In respect of denitrification, replacing the first layer of catalyst was done and a reserve layer catalyst was put into place. An acoustic sonic sootblowing system of the reserve layer was installed as part of the whole configuration. As to PM removal, low-temperature economizers were added into place. The original ESP-FF were upgraded for higher efficiency, of which particle preremoval devices and flue gas distribution devices were added. Also, the previous electric zones were divided into smaller subareas, the original local power frequency supplies were replaced with three-phase high-efficiency power supplies, and the original top electromagnetic rapping was changed into a side mechanical rapping, with their damaged plates and wires being replaced. In respect to desulfurization, a new spray absorber was established with five spray layers and one layer of alloy tray. The original liquid column absorber was removed, and each layer of spray layer had a slurry circulating pump corresponding to it. High-efficiency demisters of the three-stage ridge roof type and a six-stage demister flushing water were also installed. The original oxidation Roots blowers were removed, two high-efficiency centrifugal oxidation blowers were added to each unit,

Table 7.8 Costs of ultra-low emission engineering projects in Hequ Power Plant

	Item	Unit	Amount
1	Engineering project total static investment	million CNY	328.29
2	Annual operating hours	h	5500
3	Depreciation period	year	15
4	Installed capacity	MW	1200
5	Electricity price	CNY/(kW h)	0.3772
6	Catalyst (incremental)	million CNY	380.8
7	Liquid ammonia (incremental)	million CNY	1.57
8	Limestone (incremental)	million CNY	2.15
9	Electricity consumption (incremental)	million CNY	11.50
10	Water consumption (incremental)	million CNY	2.75
11	Maintenance and repair costs (incremental)	million CNY	4.92
12	Depreciation costs (increment)	million CNY	21.20
13	Other costs	million CNY	6.36
14	Accounting expenses	million CNY	8.31
15	Sewage subsidies	million CNY	6.14
16	Total cost incremental	million CNY	56.43
17	Newly added unit cost	CNY/kW	27.4
18	Added value for unit electricity costs	10^{-2} CNY/kW h	0.855

and eight agitators per absorber were added. The original gypsum slurry discharge pump was removed, two new gypsum discharge pumps were added, and a new powder chamber with a limestone slurry tank and three-disc dewatering machines were added. The anticorrosion upgrade to the stack was carried out. A new inner exhaust cylinder was installed inside the concrete cylinder wall of the original stack, which referred to the installation of an inner cylinder of titanium steel composite plate.

Total static investment of ultra-low emission was 328.29 million CNY, including 137.64 million CNY for desulfurization upgrade, 20.18 million CNY for denitrification upgrade, 81.35 million CNY for PM removal upgrade, 30.47 million CNY for low-temperature economizer upgrade, and 31.42 million CNY for stack anticorrosion upgrade. Based on the final audit, the new unit cost was 27.4 CNY/kW, and the added value of the unit cost was 0.855×10^{-2} CNY/kW h, as shown in Table 7.8.

7.3 Jiaozuo Power Plant

As to the ultra-low emission upgrade for the 660-MW ultra-supercritical units in Jiaozuo Power Plant, no relative upgrade was performed on either the denitration system or the ESP, only refurbishment was done to the demister, oxidizing air pipe network and desulfurization exit CEMS, to meet the ultra-low emission requirements. There was no change regarding the demister pressure difference before and after demister replacement. And accordingly, no other costs, such as power consumption and water consumption, increased. The investment for equipment replacement was 17 million CNY, and the newly added cost per unit was 12.9 CNY/kW. After the ultra-low emission upgrade, the material consumption was increased, mainly because of the increase in the consumption of limestone and liquid ammonia. The data comparison before and after ultra-low emission upgrade is shown in Table 7.9. Before upgrade, the limestone consumption in the first half of 2016 was 4.45 g/kW h, and the liquid ammonia consumption was 0.16 g/kW h. After the upgrade, the limestone consumption in the second half of 2016 was 9.13 g/kW h, and the liquid ammonia consumption was 0.24 g/kW h. If the annual electricity consumption was estimated as 6 billion kW·h, the limestone consumption would increase by 28,000 tons annually and the consumption of liquid ammonia would increase by 480 tons annually. If the price of limestone was estimated at 113 CNY/ton, and that of the liquid ammonia was estimated as 1998 CNY/ton, the power generation cost would increase annually by 4.13 million CNY.

Some researchers also compared ultra-low emission upgrade costs between coal-fired power units and gas turbines. In a comprehensive way, Wang Shumin took into consideration the differences between

Table 7.9 Data comparison before and after ultra-low emission upgraded

	Item	Unit	Amount
1	Equipment investment	million CNY	17.00
2	Limestone consumption before upgrade	g/kW h	4.45
3	Liquid ammonia consumption before upgrade	g/kW h	0.16
4	Limestone consumption after upgrade	g/kW h	9.13
5	Liquid ammonia consumption after upgrade	g/kW h	0.24
6	Power generation	MW h/year	6,000,000
7	Increased limestone	Ton/year	28,080
8	Increased liquid ammonia	Ton/year	480
9	Limestone price	CNY/ton	113
10	Liquid ammonia price	CNY/ton	1998
11	Increased limestone cost	million CNY	3.17
12	Increased liquid ammonia cost	million CNY	0.96
13	Increased total cost after upgraded	million CNY	4.13
14	Increased unit cost	CNY/kW	12.9

Table 7.10 Price comparison between natural gas-fired and near-zero emission coal-fired power plants in Shenhua Group's Beijing-Tianjin-Hebei region

Item	Unit	Beijing-Tianjin-Hebei 22 coal-fired power units' average	Beijing natural gas thermal power plant	Zhejiang Yuyao natural gas power plant
Capacity	MW	450	950	780
Investment costs	CNY/kW	3955	3532	2836
Maintenance costs	CNY/kW a	65	89	84.41
Total cost of electricity sales	CNY/kW h	0.24	0.55	0.71
Fuel costs	CNY/kW h	0.16	0.42	0.53
Electricity sales price	CNY/kW h	0.365	0.78	0.815

coal-fired power plants and natural gas-fired power plants, in terms of construction cost, maintenance cost, environmental protection investment, electricity sales cost, and electricity sales price. And he drew the conclusion that, in electricity generation, near-zero emission coal-fired power plants were better in cost effectiveness than natural gas-fired power plants [10]. For details, please see Table 7.10.

References

[1] D.Y. Zhao, Y.N. Jin, S.Q. Zhang, Cost-effectiveness analysis of pollution emission reduction measures and ultra-low emission policies for coal-fired power plants, China Environ. Sci. 36 (9) (2016) 2841–2848.

[2] Y.B. Chen, Y. Zhang, W. Ling, Practice on ultra-low emission and energy efficient technologies in coal-fired power plants, Front. Eng. Manag. 3 (4) (2016) 377–383.

[3] S.M. Wang, Practice of Near-Zero Emission and Energy Saving Upgrading in Sanhe Coal-Fired Power Plant, China Power Electric Press, Beijing, 2016.

[4] H. Xu, X.J. Yin, Economic analysis on super low emission renovation technology at large scale coal-fired power plant, Shanghai Energy 2 (2018) 131–135.

[5] S.M. Wang, J.Z. Liu, Investigation of near-zero air pollutant emission characteristics from coal-fired power plants, Proc. CSEE 36 (22) (2016) 6140–6148.

[6] Y.H. Xiong, Study on the drive of ultra-low emission and its developing direction, Environ. Prot. 42 (22) (2014) 33–35.

[7] D.H. Zhang, Y. Zhuang, R.R. Zhu, Ultra-low air pollutant control technologies for coal-fired flue gas and its economic analysis, Electric Power Construct. 36 (5) (2015) 125–130.

[8] X.X. Cheng, B.H. Wu, Y.J. Zhou, G.P. Niu, X.H. Li, D.W. Sun, Ultra-low emission reform scheme for coal-fired power plants and its economic analysis, Thermal Power Gener. 46 (11) (2017) 97–102.

[9] X.X. Cheng, B.H. Wu, Discussion on ultra-low emission technology and its economy of coal-fired power plants, in: Proceedings of the 20th Symposium on SO_2, NOx, $PM_{2.5}$ and Hg Pollution Control Technology, May 7–8, Beijing, 2016.
[10] S.M. Wang, Research on Coal-Fired Power Plants Near-Zero Emission Technologies and Applications, North China Electric Power University, 2016.

Appendix

Table A.1 Statistical of ultra-low emission coal-fired power plants

	Power generation companies	Province	Power plant	Unit	Capacity (MW)	Acceptance test	Pollutant (mg/m^3)		
							PM	SO_2	NO_x
1		Tianjin	Dagang	1	328.5	2014.2	2.8	13	32
2		Tianjin	Dagang	2	328.5	2013.11	3.33	18	33.67
3		Tianjin	Dagang	3	328.5	2014.5	3.53	9.33	33.33
4		Tianjin	Dagang	4	328.5	2014.3	2.97	17	26.67
5		Tianjin	Panshan	1	530	2015.5	2.36	4.71	36.77
6		Tianjin	Panshan	2	530	2015.12	2.1	4.65	34.48
7	Shenhua	Hebei	Sanhe	1	350	2014.7	5	9	35
8	Group	Hebei	Sanhe	2	350	2014.11	3	10	25
9		Hebei	Sanhe	3	300	2015.11	2	12	22
11		Hebei	Sanhe	4	300	2015.7	0.23	5.9	20
12		Hebei	Dingzhou	1	600	2015.12	0.74	8	22
13		Hebei	Dingzhou	2	600	2016.3	0.9	9	29
14		Hebei	Dingzhou	3	660	2014.12	2	6	17
15		Hebei	Dingzhou	4	660	2015.1	2	7	21
16		Hebei	Cangdong	1	600	2016.2	3	9	23
17		Hebei	Cangdong	2	600	2016.3	2	3	20
18		Hebei	Cangdong	3	660	2015.11	4	15	26
19		Hebei	Cangdong	4	660	2015.10	2	10	19
20		Hebei	Qinhuangdao	1	215	2015.7	2.23	8.16	26.5
21		Hebei	Qinhuangdao	2	215	2015.7	2.23	8.16	26.5
22		Hebei	Qinhuangdao	3	320	2014.12	1.5	*	8
23		Hebei	Qinhuangdao	4	320	2015.4	1.6	3	24
24		Inner Mongolia	Hubei	1	600	2016.8	4.9	<15	25
25		Inner Mongolia	Hubei	2	600	2016.8	5	15	20

26	Inner Mongolia	Zhungeer	3	330	2015.9	4.1	21	10.6
27	Inner Mongolia	Zhungeer	4	330	2015.11	3	10.23	25.05
28	Ningxia	Yuanyanghu	2	660	2014.8	4.5	30.4	24.8
29	Ningxia	Ningdong	1	330	2015.11	4.45	21	30.9
30	Ningxia	Ningdong	2	330	2015.11	3.8	29.2	41.2
31	Liaoning	Suizhong	1	880	2015.2	5	16	27
32	Liaoning	Suizhong	2	880	2014.8	5	30	44
33	Liaoning	Suizhong	3	1000	2015.9	4.42	19	46
34	Shanxi	Hequ	3	600	2016.1	2.4–4.7	11.2–22.6	13.4–38.8
35	Shanxi	Hequ	4	600	2016.6	1.59–4.24	17.7–23.2	12.8–39.3
36	Henan	Jiaozuo	1	660	2014.8	<5	<35	<50
37	Henan	Jiaozuo	2	660	2016.7	<5	<35	<50
38	Henan	Mengjin	1	600	2015.5	3.5	18	36
39	Henan	Mengjin	2	600	2015.4	3.77	12	40
40	Shandong	Shouguang	1	1000	2016.7	<1	2	18
41	Shandong	Shouguang	2	1000	2016.11	<1	<2	16
42	Chongqing	Wanzhou	2	1000	2015.9	2.92	19.9	32.2
43	Jiangsu	Chenjiagang	2	660	2016.6	<2.9	<10	<20
44	Jiangsu	Taicang	7	630	2015.11	0.3～0.4	3～8	15～19
45	Anhui	Maanshan	1	330	2015.12	0.99	5.7	22
46	Anhui	Maanshan	2	330	2016.1	0.91	6	18
47	Anhui	Maanshan	3	330	2015.12	1.06	21	46
48	Anhui	Maanshan	4	330	2016.6	0.98	6	32
49	Anhui	Jiuhua	1	320	2016.6	2.1	8.3	11.9
50	Anhui	Jiuhua	2	320	2015.8	2.35	3.5	13.25
51	Zhejiang	Ninghai	2	600	2016.3	1	3	27
52	Zhejiang	Ninghai	4	600	2015.12	<1	6	17

Continued

Table A.1 Statistical of ultra-low emission coal-fired power plants—cont'd

	Power generation companies	Province	Power plant	Unit	Capacity (MW)	Acceptance test	Pollutant (mg/m^3)		
							PM	SO_2	NO_x
53		Zhejiang	Zhoushan	3	300	2015.12	0.53	2.87	23.9
54		Zhejiang	Zhoushan	4	350	2014.6	2.46	2.76	19.8
55		Guagndong	Huizhou	1	330	2014.12	1.4	8	18
56		Guagndong	Huizhou	2	330	2016.1	0.8	9	18
57		Fujiang	Yanshi	5	300	2015.12	4.4	28	41
58		Hebei	Zhangjiakou	3	320	2015.12	<10	<35	<50
59		Hebei	Zhangjiakou	4	320	2015.12	<10	<35	<50
60		Hebei	Zhangjiakou	5	320	2015.12	<10	<35	<50
61		Hebei	Zhangjiakou	6	320	2015.12	<10	<35	<50
62		Hebei	Zhangjiakou	8	320	2015.12	<10	<35	<50
63	China Datang	Hebei	Qingyuan	1	330	2015.10	2.3	20.1	34.1
64	Corporation	Hebei	Qingyuan	2	330	2015.11	2.3	20.1	34.1
65		Shanxi	Lingfen	1	300	2015.12	3	9	26.8
66		Shanxi	Baqiao	2	330	2016.6	<5	<35	40
67		Henan	Sanmengxia	4	600	2016.2	<5	<35	<50
68		Jiangsu	Nanjing	1	660	2015.6	2.43	26.91	44.56
69		Jiangsu	Nanjing	2	660	2014.11	4.9	29	35
70	China	Hebei	Longshan	1	600	2015.6	<10	<35	<50
71	Guodian	Shanxi	Datong No.2	10	660	2016.5	2	18	36
72	Corporation	Shandong	Liaocheng	1	15	2015.6	<5	<35	<100
73		Shandong	Hezhe	5	300	2015.12	<5	<35	<50
74		Shandong	Hezhe	6	300	2016.5	<10	<35	<50
75		Gansu	Fanping	2	330	2015.11	7.02	20.99	73.83
76		Jiangsu	Taizhou	1	1000	2015.4	1.75	16.71	41.16
77		Tianjin	Beijiang	1	1000	2015.4	2.1	6.8	39.1
78		Tianjin	Beijiang	2	1000	2015.11	<20	<100	<100
79		Guangxi	Beibuwan	1	320	2013.12	*	*	<140

80		Guangxi	Beibuwan	2	320	2008.12	<20	<100	<100
81	State Power Investment	Inner Mongolia	Tongliao	5	600	2016.6	<5	<35	<50
82	Corporation	Henan	Yuxin	6	330	2016.7	<3.58	<16.83	<31.75
83		Henan	Pingdingshan	1	1000	2015.9	3.83–4.05	6–16.8	20.2–23.7
84		Henan	Pingdingshan	2	1000	2015.12	<10	<30	<50
85		Anhui	Pingwei	5	1000	2015.5	<10	<35	<50
86		Henan	Yaomeng	3	300	2016.6	<5	<35	<50
87	China	Hebei	Yuhua	1	300	2014.9	4	7	27
88	Huadian	Hebei	Yuhua	2	300	2014.12	4	29	29
89	Corporation	Tianjin	Junliangcheng	9	350	2014.11	3.28	<15	24
90		Tianjin	Junliangcheng	10	359	2014.11	3.59	<15	30.5
91		Anhui	Lu'an	4	660	2016.6	3.5	17	40
92		Shandong	Shiliquan	5	660	2015.7	1	13	29
93		Jiangsu	Jinling	1	1000	2015.1	1.7–2.3	1–9	10–12
94		Heilongjiang	Yichun	2	350	2015.12	<10	<35	<50
95		Hunan	Yueyang	6	600	2015.7	<10	<35	<50
96		Shandong	Rizhao	1	350	2015.11	<20	<50	<100
97	China	Shandong	Rizhao	2	350	2015.11	<30	<50	<100
98	Huaneng	Shandong	Rizhao	3	680	2015.11	<30	<50	<100
99	Group	Shandong	Rizhao	4	680	2015.11	1.8	13.5	35.5
100		Zhejiang	Changxin	1	660	2014.12	2.56	13.67	37.12
101		Zhejiang	Changxin	2	660	2014.12	2.15	17.57	14.8
102	China	Jiangsu	Nanjing	3	300	2016.4	<5	<35	<50
103	Resources (Holdings)	Jiangsu	Nanjing	6	1000	2015.12	<5	<35	<50

*No data.

Author Index

Note: Page numbers followed by *f* indicate figures and *t* indicate tables.

V

W

X

Y

Z

9780081024188